Pharmacology - Research, Safety Testing and Regulation

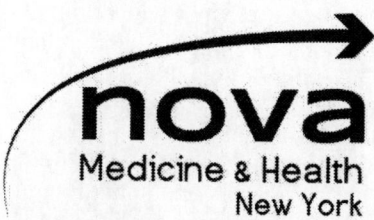

nova
Medicine & Health
New York

Pharmacology - Research, Safety Testing and Regulation

Fundamentals of Flavonoids and Their Health Benefits. A Textbook for Undergraduate, Graduate, and Postgraduate Students
Katrin Sak, M.Sc., Ph.D., Dip.Nut.Med. (Editor)
2024. ISBN: 979-8-89113-457-7 (Hardcover)
2024. ISBN: 979-8-89113-521-5 (eBook)

Time-Proof Perspectives on Bioequivalence
Carla Vitorino -Assistant Professor (Editor)
João José Sousa – Associate Professor (Editor)
António José Almeida – Professor (Editor)
Margarida Miranda (Editor)
2023. ISBN: 979-8-88697-604-5 (eBook)

Clinical Pharmacy Practices: Main Guidelines
Mohamed Emam Abdelmobdy Abdelrahim (Editor)
2023. ISBN: 979-8-88697-834-6 (Hardcover)
2023. ISBN: 979-8-88697-859-9 (eBook)

Clinical Pharmacist Tools to Deliver Rational Medication Use
Mohamed Emam Abdelmobdy Abdelrahim (Editor)
2023. ISBN: 979-8-88697-834-6 (Hardcover)
2023. ISBN: 979-8-88697-859-9 (eBook)

The Biochemical Guide to Antibiotics
David Aebisher and Dorota Bartusik-Aebisher (Editors)
2022. ISBN: 979-8-88697-193-4 (Softcover)
2022. ISBN: 979-8-88697-242-9 (eBook)

More information about this series can be found at
https://novapublishers.com/product-category/series/pharmacology-research-safety-testing-and-regulation/

Usama Ahmad
and Anas Islam
Editors

Liposomes

Advances in Research and Applications

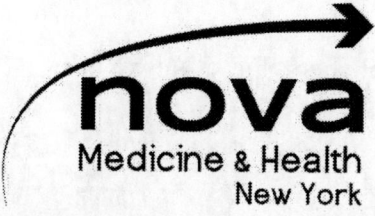

nova
Medicine & Health
New York

DOI: https://doi.org/10.52305/KGMI0837

NOTICE TO THE READER

Library of Congress Cataloging-in-Publication Data

ISBN: 979-8-89113-636-6 (softcover)
ISBN: 979-8-89113-700-4 (e-book)

Published by Nova Science Publishers, Inc. † New York

Contents

Preface

Liposomes are spherical vesicles composed of one or more lipid bilayers that can encapsulate various substances, such as drugs, genes, vaccines, and enzymes. They have emerged as versatile and powerful tool for biomedical applications, especially in the field of drug delivery. Liposomes offer several advantages over conventional drug delivery systems, such as improved solubility, stability, bioavailability, targeting, and reduced toxicity of the encapsulated agents. Liposomes have also shown great potential in the treatment of various diseases, such as cancer, infectious diseases, inflammatory diseases, and genetic disorders.

However, liposome technology is not without challenges. There are still many obstacles to overcome, such as low encapsulation efficiency, poor stability, rapid clearance, immune recognition, and limited penetration. Moreover, the development of liposome-based products requires a multidisciplinary approach, involving various aspects of chemistry, physics, biology, pharmacology, and engineering. Therefore, there is a need for a comprehensive and updated source of information that covers the fundamental principles, recent advances, and future directions of liposome research and applications.

This book, titled "Liposomes: Advances in Research and Applications", aims to fill this gap by providing a state-of-the-art overview of liposome science, from its historical origins to its current status and prospects. The book consists of five chapters, each written by experts in the field, that cover the following topics:

- Chapter 1 introduces the basic concepts of liposome formation, structure, classification, and characterization, as well as the various applications of liposomes in different fields, such as medicine, biotechnology, agriculture, and cosmetics. It also highlights the role of liposomes in enhancing the stability and delivery of bioactive compounds, such as antioxidants, vitamins, and probiotics.

- Chapter 2 reviews the recent developments and innovations in liposome technology, such as novel methods of preparation, functionalization, and characterization. It also addresses the challenges and solutions related to liposome encapsulation efficiency, stability, biodistribution, and release. It further explores the potential of liposomes in delivering various types of drugs, such as small molecules, peptides, proteins, nucleic acids, and vaccines, for the treatment of various diseases, such as cancer, diabetes, Alzheimer's, and tuberculosis.
- Chapter 3 focuses on the application of liposomes in cancer therapy, one of the most promising and active areas of liposome research. It discusses the advantages of liposomes in achieving targeted, controlled, and enhanced delivery of anticancer agents, such as chemotherapeutics, immunotherapeutics, and radiotherapeutics. It also critically analyzes the current status, challenges, and future prospects of liposome-based cancer therapy, including the clinical trials, regulatory issues, and market potential.
- Chapter 4 zooms in on the specific application of liposomes in the treatment of three major types of cancer affecting women: breast, cervical, and ovarian cancers. It emphasizes the need and potential of liposomes in revolutionizing the current cancer treatment modalities, such as surgery, chemotherapy, radiotherapy, and hormone therapy. It also discusses the existing challenges and recent strategies in developing liposome-based formulations for these cancers, such as passive and active targeting, stimuli-responsive release, and combination therapy. It further provides an overview of the ongoing clinical trials and future directions of liposome-based cancer therapy for these cancers.
- Chapter 5 explores the emerging field of gene therapy and the crucial role of liposomes as gene delivery vehicles. It describes the advantages and challenges of liposome-mediated gene delivery, such as transfection efficiency, specificity, safety, and immunogenicity. It also details the recent advancements and trends in liposome-based gene delivery systems, such as cationic liposomes, fusogenic liposomes, stealth liposomes, and multifunctional liposomes. It further highlights the potential of liposome-based gene delivery in treating various genetic disorders, such as cystic fibrosis, hemophilia, and muscular dystrophy.

The book is intended for a broad audience, including researchers, students, teachers, clinicians, and industry professionals, who are interested in learning more about liposome science and its applications. The book is written in a clear and concise manner, with ample illustrations, tables, and references, to facilitate the understanding and appreciation of the subject. The book also provides a balanced and critical perspective on the achievements, challenges, and opportunities of liposome research and applications.

The book is the result of a collaborative effort of several authors and editors, who have contributed their expertise, experience, and insights to this project. The process of compiling the book was not without difficulties, as it involved integrating multidisciplinary insights, addressing rapid advances in the field, balancing depth and accessibility, and navigating the complex landscape of cancer therapies. However, these challenges were overcome with rigorous discussions, continuous updates, and careful revisions. The editors would like to thank all the authors for their valuable contributions and cooperation. The editors would also like to acknowledge the support and guidance of Nova Science Publishers Inc.

We hope that this book will serve as a useful and informative resource for the readers, and inspire them to further explore and advance the field of liposome science and applications.

Dr. Usama Ahmad
Associate Professor
Faculty of Pharmacy
Integral University Lucknow, India

Mr. Anas Islam
Lecturer
Faculty of Pharmacy
Integral University Lucknow, India

Chapter 1

Liposomes: History, Properties, and Applications

Shivangi Sharma[1]
Anas Islam[2]
Mohd Muazzam Khan[2]
and Usama Ahmad[2,*]

[1]Institute of Pharmacy, Shri Ramswaroop Memorial University, Barabanki, Uttar Pradesh, India
[2]Faculty of Pharmacy, Integral University, Lucknow, Uttar Pradesh, India

Abstract

The use of liposomes in nanotechnology has become more widespread in recent years. Liposomes are self-assembled particles having a water chamber that is surrounded by a lipid bilayer. This chapter has covered the liposomes in great detail, including how and where they were created, a broad description of their structure, and what distinct types of lipids they are composed of. Depending on their size, liposomes can be classified into several distinct varieties, such as enormous liposomes, multilamellar liposomes, and unilamellar liposomes. Depending on the context, different liposomes can be produced using various preparation techniques. In this chapter, we have also learned about the several fields in which liposomes have been extensively employed for the treatment of various diseases, biotechnology research, cosmetic items, nutraceuticals, and food. To increase the stability of the core, liposomes are important in the encapsulation of a variety of bioactive compounds (BACs), including functional food additives. In this chapter, we have

* Corresponding Author's Email: usamaahmad.10@outlook.com.

In: Liposomes
Editors: Usama Ahmad and Anas Islam
ISBN: 979-8-89113-636-6

also discussed the future perspective and scope where liposome technology can be exploited, the different fields in which liposome nanotechnology can be a boon, and the revolution it can make in pharmaceuticals.

Keywords: liposomes, vesicles, nanotechnology, phospholipids, drug delivery system

1. Introduction

Finally, nanotechnology is fully embraced by the medication distribution industry. Intelligent drug delivery systems are continuously improved to promote therapeutic activity and minimise unrequired side effects (Safari et al., 2014). The potential of using nanoparticles as a powerful medicine delivery mechanism is enormous. Numerous investigations have been done on nanosystems with different biological properties and compositions for use in gene delivery and medicine (Suri et al., 2007).

A very important and most widely used nanotechnology used for delivering numerous drugs and achieving targeted delivery is liposomes. Liposomes are vesicles that are artificially made of one or more phospholipid bilayers containing both lipid- and water-soluble compounds. Liposomes can be optimised to fit certain medicine and targets by formulating them in a variety of sizes, bilayer melting temperatures, and surface charges. As spherical vesicles with particle sizes ranging from 30 to several micrometres, liposomes are often identifiable. Liposomes, or synthetic lipid vesicles, emerge when amphiphiles (surface-active molecules comprising a hydrophilic group and a hydrophobic group in the chain at opposing ends) are exposed to water. Amphiphiles are organised as one or more concentric bilayers (lamellae) alternating with aqueous compartments when the mass ratio of the amphiphile to water is acceptable and the temperature is also appropriate. The drugs and other solute molecules from the aqueous media are captured by liposomes as they develop. In contrast, the lipid phase may embrace lipid-soluble drugs paired with the amphiphile to form portions of the bilayers. (Gregoriadis et al., 2010).

In the pharmaceutical and cosmetic sectors, liposomes are widely employed as molecules carriers. Liposomes have gained popularity both as an experimental system and commercially due to their biodegradability, low toxicological profile, ability to bind both hydrophilic and lipophilic medications (Johnston et al., 2007), and capacity to target drug delivery to cancer tissues (Hofheinz et al., 2005). The most recent delivery method employed by medical researchers to transport medications that operate as curative promoters to the guaranteed bodily parts is called liposomal encapsulation technology (LET). This delivery system idea was aimed at providing the body with essential combinations (Akbarzadeh et al., 2013). Numerous liposome applications have been researched to date, and some are still being investigated. In many circumstances, harmful drug side effects significantly restrict the use of chemotherapy. The geographical and temporal distribution of the drug molecules contained in liposomes can be altered, potentially reducing hazardous side effects and improving therapeutic effectiveness. Most of the liposomes' medicinal uses that are now in the preclinical stage are for the treatment of cancer. Early research generally indicated that liposome-encapsulated medicine was less toxic, but most of the time, the drug molecules were not accessible, drastically impairing both effectiveness and toxicity. Studies on liposomes for the prevention and treatment of various diseases are ongoing. Examples include insulin, prostaglandins, steroid and non-steroid anti-inflammatory drugs, antivirals, and liposomal antibiotics. Pulmonary, topical, and other parenteral administration methods are being studied (Lasic, 1998). The Swiss Serum Institute in Bern, Switzerland, has successfully introduced a liposomal hepatitis vaccine (Gluck et al., 1994) The use of targeted delivery and immunotherapy are two more areas where experts anticipate advancement. However, creating formulations for the treatment of tropical antiparasitic diseases such as leishmaniasis and malaria would be much simpler (New et al., 1978). Therefore; we shall cover the many forms, relevance, and applications of liposomes in this chapter.

1.1. Historical Development of Liposomes

Considering liposomes resemble cell membranes along with the discovery of liposomes in the middle of the 1960s (Bangham et al., 1965), cell biologists now have a special tool for studying a variety of cell membrane activities, such as cell fusion, antigen presentation, and membrane pumps. However, liposomes were not taken into consideration as a potential vehicle for the delivery of therapeutically effective substances until many years later (Gregoriadis et al., 1972; Gregoriadis et al., 1972). It is common to talk about decades characterised by important turning points while talking about the delivery and targeting of drugs using liposomes. The initial understanding of the system's activity in vivo, namely its interaction with the biological environment in the living animal, and as a result, the development of several therapeutic applications, are therefore recalled during the 1970s (Gregoriadis, 2016). Additionally, at this time improvements were made to liposomal stability in biological fluids like blood, to the science behind liposomes (Gregoriadis, 1992) when it comes to developing strategies for trapping high yield and maintaining undamaged liposomes while storing them. In 1890, Lord Raleigh studied the friction at the interface between a triglyceride, castor oil, and water (Fisher et al., 1985). Later, in 1925, Gorter and Grendel made the case that phospholipid molecules, or "lipid bilayers," made up the cell membrane (Gorter et al., 1925) "Fluid Mosaic Model" in 1972 proposed by Singer, as a result of these results, and it is still widely accepted today (Singer et al., 1972).

At the Babraham Institute in Cambridge, British hematologist Dr. Alec D. Bangham FRS first characterised liposomes in 1961 (published 1964). The finding was achieved when authors Thenando, Horne, and Bangham added a negative stain to dried phospholipids to test the university's new electron microscope. Images taken under the microscope provided the basic concrete proof that the cell membrane is a bilayer lipid structure because of the striking similarities to the plasmalemma. A membraneous lipid bilayer completely encloses an aqueous volume in liposomes, which are structurally concentric bleeder vesicles (Dua et al., 2012).

Gerald Weissman from the US showed that lysozyme, an enzyme, could be trapped inside liposomes in addition to using them as models for lysosomes (Sessa et al., 1970). Brenda Ryman was researching glycogen storage disease at Harrow at the time, and it appears that she had thought that liposomes may be employed as enzyme carriers for lysosomal storage disease replacement treatment because of their apparent capacity to entrap lysozyme (Leserman, 2008). Bangham's systems were first referred to as "multilamellar smectic mesophases"; subsequently, Gerald Weissmann suggested the term "liposomes" to describe them. When combined with other findings, Bangham's 1970s hypothesis that "something like liposomes must have been available to house the first forms of cellular life" had a significant influence on evolutionary history (Deamer, 2010).

2. Liposome Formation and Structure

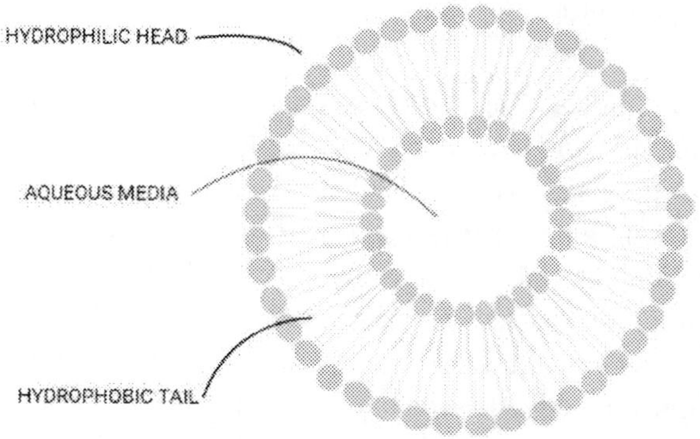

Figure 1. Structure of liposome.

The globular lipid bilayers known as liposomes, which range in size from 50 to 1000 nm, are useful delivery systems for substances with biological activity. The components of liposomes include both structural and nonstructural elements. The principal structural components include:

- Natural lipids: Although there are many structural variants among lipids, they always feature a hydrocarbon tail, as well as hydrophilic and hydrophobic head groups. A net neutral charge, a negative charge, a positive charge, or a zwitter ion (both a positive and negative charge) might be present in the headgroup. It is also possible to chemically alter the headgroup to allow conjugation with other molecules (Large et al., 2021).
- Phospholipids: Sphingolipids and phosphoglycerides are the two types of phospholipids that make up the majority of structural elements of biological coverings. The most common phospholipid is the phosphatidylcholine (PC) molecule; the PC molecule is the most common type of phospholipid. The phosphatidylcholine molecule, which is water insoluble and aqueous conditions, closely arranges itself in planar bilayer sheets to minimise the negative contact between the bulk aqueous phase and the long hydrocarbon fatty series. Most liposome formulations contain glycerols, including phospholipids, which represent more than 50-52% of the weight of lipids in biological membranes. These are derivatives of phosphatidic acid (Daraee et al., 2016).

The commonly used phospholipids are Phosphatidylcholine, Phosphatidyl serine (PS), Phosphatidyl ethanolamine, Phosphatidyl inositol (PI), and Phosphatidyl Glycerol (PG).

- Sphingolipids: Sphingolipids are an additional natural lipid class that exists predominately in the membranes of mammalian cells. Sphingolipids, which have an 18-carbon amino acid sphingosine backbone and are essential for cell membrane formation and regulatory signaling molecules, are lipids (Cuperlovic-Culf, 2013). Sphingolipids, such as sphingomyelin (SM), have been shown to boost therapeutic effectiveness by increasing stability in vivo and lengthening blood circulation. Sphingolipids called ceramides are made up of a single fatty acid and sphingosine liposomal formulations are incorporated with ceramides to enhance the fluid nature of the membrane & the potential of cutaneous cells to unite (Tokudome et al., 2009).
- Cholesterol: High levels of cholesterol, cholesterol to phosphatidylcholine, can exist in phospholipid membranes without generating a bilayer structure. In the membrane, cholesterol is located in the centre of the bilayer, parallel to the acyl chains, with its hydroxyl group facing the aqueous surface. Although interactions between definite head groups and hydrophobic molecules have both been linked to the high solubility of cholesterol in phospholipid liposomes, it remains unknown how cholesterol is organised in the bilayer (Daraee et al., 2016).
- Surfactants: Surfactants are categorised as compounds that lower surface tension in a liquid when introduced. Surfactants are helpful ingredients in liposome compositions and are also referred to as edge activators. Edge activators, which are generally single-acyl-chain surfactants, increase the deformability of arteries by thinning the lipid bilayer of liposomal nanoparticles (Chen et al., 2013). Edge activators have been shown to enhance the ability of liposomes used in anticancer (Dorrani et al., 2016; Zeb et al., 2016), antifungal (Perez et al., 2016), and transdermal applications to enter the dermis (El Zaafarany et al., 2010). The following substances are frequently used as edge activators: sodium cholate, Span 60, Span 80, Tween 60, and Tween 80. Furthermore, the charge of edge activators can be used to improve treatment effectiveness (El Maghraby et al., 2004).

Table 1. Advantages and disadvantages of liposomes

Liposome Advantages	Liposome Disadvantages
They are adaptable, nontoxic, totally biodegradable, biocompatible, and nonimmunogenic for systemic and non-systemic therapy.	Fusion of encapsulated drug/Molecules
They help reduce the amount of hazardous medication exposure to hazardous medications in delicate tissues.	Low solubility
flexibility to combine with ligands that target a specific spot to produce active targeting	Sometimes, phospholipids undergo oxidation. and hydrolysis-like reaction
Site avoidance effect	Production cost is high
Liposome encapsulation improved stability.	Fewer stables
Liposomes reduce the toxicity of the encapsulated agent.	Short half-life

3. Method of Liposome Preparation

There are four standard phases for all liposome preparation techniques:

1. Lipids from organic solvents are dried.
2. Dispersion of lipids in a watery medium.
3. Cleaning of the final liposome.
4. Examining the finished item.

3.1. Thin Film Hydration

The method that produces liposomes most often is thin film hydration. After being solubilised, lipids and amphiphilic molecules are mixed in an organic solvent. A thin layer of lipids is then left over after the solvent is extracted from the mixture using a rotary evaporator while it is under vacuum. The mixture is then added to a flask with a round bottom. The thin film is then hydrated in a solution that may also include one or more hydrophilic drugs for encapsulation. It is a widely used technique that is simple to use. The drawback of this approach is the abundance of heterogeneous MLVs that the phospholipids in the water buffer create (Liu et al., 2022).

According to Elsan et al., 2019 preparation and evaluation of new cationic gene-based liposomes with cyclodextrin synthesized by thin film hydration and microfluidic systems, the efficiency of lipoplexes was improved by both techniques when carboxymethyl-cyclodextrin was used (Elsana et al., 2019). Similar to how Tagrida et al., 2021 prepared liposomes loaded with betel leaf ethanolic extract (BLEE) by the thin film hydration method, many other liposome formulations were created by this method. The results indicated that the betel leaf ethanolic extract loaded liposome could be an effective system to enhance the antioxidant activities of the extract.

3.2. Method of dispersion of Solvent

3.2.1. Ether Injection Method

A solution of lipids dissolved in dicthyl ether or an ether-methanol mixture is gradually added to an aqueous solution of the object to be encapsulated at a temperature between 55° and 65°C or under reduced pressure. Liposomes are created when ether is removed under vacuum. The population's heterogeneity (70 to 220 nm) & the compounds that must be encapsulated's exposure to organic solvents at high temperatures are the technique's main limitations (Deamer et al., 1976; Schieren et al., 1978).

This strategy has many disadvantages, such as extremely poor hydrophilic chemical encapsulation efficacy, very poor lipid solubility in ethanol, and low lipid concentrations in the final solution due to the high ethanol content. Additionally, the ethanol concentration should not exceed 7.5% to prevent liposome destabilisation, which lowers the maximum quantity of lipids that are permitted (Jaafar-Maalej et al., 2010; Pons et al., 1993).

Justo et al., 2011 examined the effects of process variables on the properties of liposomes made by injecting ethanol The findings show that a reasonably narrow distribution of tiny unilamellar vesicles may be obtained using the ethanol injection approach, modifying the experimental conditions as needed. As a result, the goal of creating vesicles with the correct diameter and size distribution was mostly achieved (Justo et al., 2011). Charcosset et al., 2015 prepared liposomes using the ethanol injection method on a large scale and discovered that the technique appears to be suitable for industrial production due to its scaling properties and a broad range of possible operating conditions.

3.2.2. Ether Injection

An ethanol-lipid solution is swiftly injected into a sizeable excess of buffer. MLVs develop immediately. Because ethanol forms an azeotrope with water, the population is heterogeneous (30 to 110 nm), liposomes are very diluted, and there is a high likelihood that various biologically active macromolecules will inactivate in the presence of even small amounts of ethanol; it is difficult to completely remove all of the ethanol from a sample (Batrzi et al., 1973). This approach has certain drawbacks as well. For example, the water and ether phases must be at different temperatures throughout the injection process, and ether may interfere with the encapsulation of some substances. It is recommended to inject the lipid suspension into the water/buffer under vacuum and at a slower rate than when using the ethanol injection technique. However, the liposomes created with this technique have better encapsulation rates. In contrast to the ethanol injection approach, LUVs rather than SUVs are created (Torchilin et al., 2003). Immunostimulating Complexes (ISCOM) were generated by Pham et al., 2006, and the results demonstrated that ethyl injection significantly outperforms several of the drawbacks of the currently used ISCOM synthesis techniques, which are frequently time-consuming and frequently require extra surfactant.

3.3. Reverse Phase Evaporation Method

This technique improved liposome technology since it made it possible to create liposomes with a high aqueous space-to-lipid ratio and the capacity to entrap a sizable amount of the supplied aqueous material for the first time. The formation of inverted micelles is the foundation of reverse-phase evaporation. These inverted micelles are created using sonication to mix an organic phase, which solubilises the amphiphilic molecules, and a buffered aqueous phase, which includes the water-soluble molecules that will be contained in the liposomes. These inverted micelles grow thick and resemble a gel when the organic solvent slowly evaporates. Reverse phase evaporation-produced liposomes can be made from a range of lipid formulations and have a higher aqueous volume-to-lipid ratio than hand-shaken or multilamellar liposomes (Akbarzadeh et al., 2013; Kataria et al., 2011). Imura et al., 2003 created soybean lecithin liposomes utilising the supercritical reverse phase evaporation method, and the results demonstrated that the technique helped create inexpensive, mass-produced liposomes composed of lecithins. (Chen et al., 2013) used reverse phase evaporation to create metronidazole liposomes, and the findings showed that this process was successful in producing liposomes with an average volume diameter of 190–350 nm and a coating ratio of up to 48.2% (Chen et al., 2013).

3.4. French Pressure Cell: Extrusion

The extrusion method involves forcing a liposome suspension through a film filter with predetermined pore sizes (Olson et al., 1979). To complete the extrusion process, the extrusion technique needs an extruder equipped with a pump that forces suspensions through the films. A typical technique for sizing liposomes is extrusion (Hunter et al., 1998). The extrusion procedure is straightforward, quick, gentle and repeatable. It can create uniform, regulated liposomes of average size. By sequentially extruding down through a 0.2-m membrane. To maintain an acceptable degree of aqueous phase encapsulation, Olson et al., 1979 engineered liposomes with a predetermined size distribution. The results showed that the resultant vesicles show a relatively uniform size distribution with a mean diameter of 0.27 m (Imura et al., 2003).

3.5. Sonication

One of the most popular techniques for prepping SUVs is sonication. The cohabitation of MLVs and SUVs, the ineffectiveness of encapsulation, and contaminated probe tips are the main drawbacks of this approach. Both probe sonication and bath sonication are sonication techniques (Riaz et al., 1996).

Probe sonication-The tip quickly engulfs the liposome dispersion. This method significantly increases the amount of energy that goes into the lipid dispersion process. The vessel needs to be submerged in a water/ice bath because the coupling of energy at the tip results in localised heat. More than 5% of the lipids can be deesterified for up to an hour during the sonication process. Titanium will also flake off while using the probe sonicator and contaminate the fluid.

Bath sonication- A bath sonicator is used to sonicate the cylinder containing the liposome dispersion. In comparison to direct dispersal, which uses the tip during sonication, this approach typically makes it simpler to control the temperature of the lipid dispersion. In contrast to probe units, the material that is sonicated can be kept safe in a sterile vessel or in an inert environment (Kataria et al., 2011). High-encapsulation liposomes were created thanks to the sonication process used by Perrett et al., to make liposomes (Perrett et al., 1991).

3.6. Stability and Surface Charge

Physical, chemical, and biological lipid stability are the three different types. When discussing how well the liposomal formulation can hold onto its qualities over time, the terms "physical" and "chemical" stability are frequently used. The long-term stability may be affected by phospholipids' vulnerability to various chemical degradation processes, such as ester hydrolysis links and peroxidation of unsaturated acyl chains (Yadav et al., 2011).

Because of the chemistry of their head groups and the pH of the solution, each phospholipid displays a net charge. The liposome compositions influence the liposome's overall zeta potential of the liposome, which has an impact on how well the therapeutic delivery properties of the vesicles. Lipids containing negatively charged head groups make up anionic liposomes or those with negative surface charge. Anionic liposomes are used for DNA transfection (Patil et al., 2004).

Negatively charged lipids were added to liposomal formulations to improve the distribution of cardiotoxic drugs such as doxorubicin and lessen systemic toxicity compared to free drug administration (Forssen et al.,1981). Positively charged lipids can be used to create liposomes with a positive surface charge. The most commonly utilised lipid in cationic liposomes includes 1,2-dioleoyl-3-trimethylammonium-propane (DOTAP) It is well known that cationic liposomes are effective in delivering siRNA and nonviral gene transfection (Clark et al., 2000; Guo et al., 2014).

4. Types of Liposomes

There are different classifications of liposomes, based on size they are classified into two types.

4.1. Unilamellar liposomes (UV)

They are spherical vesicles that are contained by a single amphiphilic lipid bilayer or a group of similar lipids. Additionally, they have an aqueous solution in the centre of the vesicle. Using the freeze-drying method, Wang et al., 2006 produced unilamellar liposomes, and the results demonstrated that they had a respectably huge encapsulation efficiency & remarkable stability during the long run. Kirby et al., 1980 conducted in vivo and in vitro studies on the stability of small unilamellar liposomes. They found that after intravenous infusion, the shelf liposomes retained their unilamellar structure and were able to remain neutral in blood. They have also shown that liposomes can develop for efficient use in systems both *in vivo* and *in vitro* when the cholesterol content is adjusted to increase or decrease their stability.

4.2. Multilamellar Liposomes

They are made up of multiple unilamellar vesicles that develop in the interior of smaller unilamellar vesicles, giving them an onion-like shape. As a result, it develops a multi-lamellar structure made up of layers of water that separate concentric phospholipid spheres. The co-encapsulation of curcumin and vitamin D3 by multi-lamellar liposomes was created by Chaves et al., 2018 and the results showed that it is possible and that high content of bioactives was maintained during storage duration. Joo et al., 2013 have shown that this method of packaging may offer a new treatment option for cancer and other diseases. They generated crosslinked multilamellar liposomes for the controlled delivery of anticancer drugs. The results of Detoni et al.,2009. In the preparation of multilamellar liposomes containing essential oils for Zanthoxylum tingoassuiba, useful antimicrobial properties are capable of being incorporated in appreciable amounts into prepared vesicular dispersions, and the appropriate size is also obtained by a thin-film hydration method.

4.3. Small Unilamellar Vesicles (SUVs)

The creation of a single membrane or unilamellar liposomes can be done using various techniques. These systems cannot provide prolonged supply times of multimembrane preparations. Single-walled vesicles, on the other hand, may have additional benefits, such as increased peptides in the blood, long shelf life, better absorption, etc (Weiner et al., 1987).

De et al., have made drug-loaded liposomes to evaluate whether small unilamellar liposomes may be useful in the treatment of COPD. The results suggest that liposomes may have a beneficial therapeutic effect on chronic respiratory diseases, and the method also allows preparations of SUVs approximately 50 nm in length (De et al., 2018). The biocompatibility of the liposomal formulation, its capacity as a drug carrier, and stability were considered respectable. Caffeine Encapsulated Small Unilamellar Liposomes were created by Chorilli et al., who discovered that the liposomes worked well as slow-release vehicles for caffeine (Chorilli et al., 2013).

4.4. Large Unilamellar Vesicles (LUVs)

LUVs are a great model for natural membranes due to their size range of more than 100 nm. They may also be used to research membrane proteins. Because prolonged circulation times have now been achieved for large unilamellar liposomes with a large, entrapped volume and effective drug capture characteristics, large unilamellar liposomes for absorption into the reticular endothelial system (Allen et al., 1987). The finding suggested the possibility of using these altered liposomes to target therapeutic agents in tissues. Rodrigues et al., created large unilamellar liposomes to study the interaction between rifampicin and isoniazid. Their research led to the conclusion that stable preparations of liposomes containing both drugs could be made and used to treat tuberculosis (Rodrigues et al., 2003). By incorporating bacteriorhodopsin into large unilamellar liposomes using reverse phase evaporation, Rigaud et al., demonstrated that, in addition to the homogeneity of the vesicles' sizes, their relative impermeability to ions, and the bacteriorhodopsin's favourable orientation, the large size of the proteoliposomes has many benefits (Rigaud et al., 1983).

4.5. Giant unilamellar vesicles (GUVs)

To investigate the mechanical characteristics of lipid bilayers, giant liposomes larger than 10 um in diameter have been used. Giant liposomes must be generated in an ionic state that is similar to the physiological one and have their unilamellarity established before they can be used as a cell model.The results of Akashi et al., 1996 study of gigantic unilamellar vesicles under an optical microscope revealed that the huge unilamellar liposomes made using their approach are flaccid and may adopt a variety of forms.

Table 2. Types of liposomes

S. No	Vesicle type	Abbreviation	Diameter size	Comment	Reference
1.	Unilamellar liposome	UV	All size range	The long-term storage process was claimed to have outstanding stability and high encapsulation efficiency.	Wang et al., 2006
2.	Multilamellar liposome	MLV	More than 0.5µm	They were found to be high-content retainers for bioactives: turmeric and vitamin D_3. Cross-linked multilamellar liposomes were found to be efficient for the controlled delivery of anti-cancer drugs.	Chaves et al., 2018; Joo et al., 2013
3	Small unilamellar vesicles	SUV	20-100nm	For the treatment of chronic respiratory illnesses, there is good therapeutic promise. caffeine transporters have been shown to be efficient, likely operating as slow release vehicles.	De et al., 2018; Chorilli et al., 2013
4	Large unilamellar vesicles	LUV	More than 100nm	They have a large, entrapped volume and efficient drug-capture characteristics. In the treatment of tuberculosis, liposomes made with isoniazid and rifampicin were used.	Allen et al., 1987; Rodrigues et al., 2003
5	Giant unilamellar vesicles	GUV	More than 0.1µm	Incorporation in GUV membrane preserves Membrane protein	Biner et al., 2016
6	Stealth liposomes (PEGylated liposomes)		<200nm	Their formulations largely increased the solubility & bioavailability of ursolic acid. They were also found to exert a higher antitumor activity.	Yang et al., 2014. Cosco et al.,2009

In their work, Biner et al., 2016 used charge-mediated fusion to transfer membrane protein to large unilamellar liposomes. The outcome revealed that the GUVs ensured to retain the orientation of the membrane protein after incorporation into the GUV membrane.

The gentle hydration of phosphatidylcholine films doped with sugar by Tusmoto et al., 2009 resulted in the preparation of giant unilameller liposomes; this method can be preferred for preparing giant liposomes, particularly in the life sciences, due to its ease and benign conditions. Moreover, sugar's ability to shield lipid membranes was well recognised.

4.6. Stealth Liposomes

PEGylated Liposomes (Polyethylene glycol, or PEG), among the many polymers explored to speed up the blood circulation time of liposomes, have been a popular choice. There are several techniques to incorporate the polymer into the vesicle surface, and a cross-linked lipid (PEG distearoyl phosphatidylethanolamine) is used to join the polymer in the liposome membrane. This method is now the most prevalent. Several methods are used for physical adsorption of the polymer on the surface of liposomes; incorporating PEGylated conjugates during liposome synthesis, or covalently attaching reactants to surfaces of liposomes that have previously been formed (Immordino et al., 2006).

To study the anticancer effects of foliate-targeted ursoliolic acid, Yang et al., 2014 have discovered a stealth liposome that has not been studied in vitro or for human use. The results have shown a much higher increase in the solubility and bioavailability of the ursolinol in this formulation. The fatty acid stabilised urea liposome that is aimed at folic acid provides the potential to cure human epidermoid carcinoma tumour cells, according to research on cytotoxicity, apoptosis test, pharmacokinetics, and plasma stability.

To treat pancreatic cancer, Cosco et al., 2009 developed gemcitabine-loaded PEGylated tiny unilamellar liposomes. The results revealed that these liposomes are effective against this disease and had stronger antitumor activity.

5. Liposome Encapsulation Techniques

5.1. Encapsulation of Hydrophilic Drugs

Hydrophilic drugs are hydrated when they are encapsulated with lipids. Drugs can reach the liposome core with this technique, while other materials stay on the liposome's outside. The remaining components will release the medication from the liposome. These two components (drugs and remaining outside materials) can be purified using dialysis, gel filtering, and column chromatography. Techniques for dehydrating and rehydrating can be used to obtain high protein and DNA encapsulation (Chatterjee et al., 2012). According to a study by Jaffar et al., 2010, a hydrophilic drug (cytarabine; Ara-C) and a lipophilic drug (beclomethasone dipropionate; BDP) have been effectively entrapped in small liposomes by employing the ethanol injection technique.

5.2. Encapsulation of Hydrophobic Drugs

The hydrophobic drug-encapsulating area of liposomes is the phospholipid bilayer. The transport of these medications (such as verteporfin) toward the inner and outer aqueous regions of liposomes would be reduced by their trapping. These medications are encapsulated using phospholipids and organic solvents to solubilise the medications. The hydrophobic region of the liposome is where the drug is trapped. After that, laser light can be used to activate a drug for the purpose of treating wet macular degeneration (Group1A, 2001).

For the inner watery core, Okamoto et al., 2018 created albumin-encapsulated liposomes to encapsulate hydrophobic medicines. The findings demonstrated that hydrophobic compounds may be loaded more readily into liposomes' inner watery cores of liposomes because of albumin encapsulation, without the use of extra chemicals or laborious techniques.

Chen et al., 2007 developed liposome-encapsulated cyclodextrin inclusion complexes for Hydrophobic Drugs. The results showed that the hydrophobic drug inclusion complexes were captured in liposomes by the dehydration-rehydration method. Compared to regular liposomes, inclusion complexes improved the loading efficiency of Indomethacin while also altering the in vitro release.

5.3. Active Loading and Passive Encapsulation Methods

Drug encapsulation using the active loading method reduces the permeability to membranes. For example, a transmembrane gradient in combination with a pH gradient is employed to push drug molecules into empty vesicles. Ammonium sulfate or citrate buffer was used to make pH gradients, which were then found to increase the effectiveness of amphiphilic drug encapsulation while maintaining a simple, low-cost and safe process (Tanzina et al., 2011). These techniques are used for commercial drug loading of numerous FDA-approved liposomal systems, including Doxil, MyocetTM and DuanoXome. The physicochemical properties of liposomes, such as their size, surface charge, composition, and surface changes, may affect the preparation's ability to encapsulate (He et al., 2019).

6. Liposome Applications in Medicine

In the last 30 years, there has been significant advancement in liposome research. It is now possible to create a wide range of liposomes with different sizes, phospholipid contents, cholesterol constituents, and surface morphologies that are suitable for several applications (Mayer et al., 1990).

6.1. Drug Delivery Systems

Typically, liposomes are used to treat various respiratory ailments. Liposomal aerosols can be designed to provide extended release, prevent local inflammation, reduce adverse effects, and improve stability throughout the broad aqueous core. There are currently many injectable liposome-based medications on the market, including ambisome, fungisome, and Myocet. Lipid composition, charge, size, drug & lipid ratio, and delivery methods all play a role in how well liposomal drugs are delivered to the lungs. The newly discovered use of liposomes to deliver DNA to the lungs suggests that our understanding of how to use them to administer macromolecules by inhalation is currently advancing. The formation of proteins using liposomes can be improved with the use of this new knowledge. For liposome inhalation, the liquid or dry form is used, and nebulisation releases the drug.

Drug powder liposomes have been created by milling or spray drying (Farooque et al., 2021).

Both the anterior and posterior segments of the eye have long been treated using liposomes. Some eye conditions include dryness, keratitis, proliferative vitreoretinopathy, endopthelmitis, and rejection of a corneal transplant. Retinal abnormalities are the main cause of blindness in underdeveloped countries. Both a genetic transfection vector and a carrier for monoclonal antibodies are made of liposomes. The use of heat-activated liposomes in focused lasers as well as the heat-induced release of liposomal therapeutics and dyes for targeted delivery are recent therapeutic methods for the treatment of targeted tumours and neovascular artery occlusion, angiography, retinal and choroidal blood vessel stasis, and other conditions. Up to now, two patent medicines for liposomal medicinal compositions have been approved, while several others are the focus of ongoing research. A liposomal drug for ocular use called "Verteporfin" has corporate support. The liposome will be important for the therapeutic, diagnostic, and research aspects of ophthalmology in the future (Abra et al., 1980).

6.2. Targeted Therapy and Site-Specific Drug Delivery

Due to their biocompatibility and biodegradability, liposomes have recently become a method of delivering drugs to the brain (Mc Caluey, 1992). Small (less than 100 nm) or even large liposomes can easily flow through the blood–brain barrier (BBB). On the other hand, SUVs associated with cognitive drug carriers can penetrate the BBB by transcytosis mediated by receptors or absorptive mechanisms. Cells endocytose cationic liposomes by absorptive mechanisms, although it is unknown whether the BBB is crossed by absorptive-induced transcytosis. Mannose-coated liposomes are transported to the brain and aid in the transport of drugs over the BBB. Leu-enkephalin, met-enkephalin kyotorphin, and neutropeptides often do not cross the blood-brain barrier when administered systemically. Due to the adaptability of this method, the antidepressant amitriptyline typically passes through the BBB (Schroeder et al.,1998). It has been demonstrated that liposomes that are antibody-targeted can enhance the specific toxicity of liposomal anticancer medicines to grown cells (Heath et al., 1983). However, antibody-targeted liposomes were quickly removed from the bloodstream, which prevented them from reaching tissues other than MPS in vivo (Papahadjopoulos et al., 1987). Following the development of long-circulating (PEGylated) liposomes, it was found that, despite some accumulation of these liposomes being visible at target sites easily accessible from the circulation when antibodies were attached to the liposome surface, their antigen binding was obscured by the presence of PEG in the same liposomes, particularly longer chain PEG molecules (Klibanov et al.,1991; Mori et al., 1991).

6.3. Liposomal Vaccines

After being injected intramuscularly, liposomal immune adjuvants act by slowly and passively releasing encapsulated antigens into the targeted lymph node. By using phosphatidyl serine to target liposomes, lymphoid tissue can accumulate liposomes. By immunising bacteria, soluble antigen and deoxynivalenol, one can create a liposomal vaccination (Watson et al., 2012).

Inflexal V, an inactivated trivalent influenza vaccine, has liposomes with viral antigens on its surfaces that are 150 nm in size (Bulbake et al., 2017; Mischler et al., 2002; Daeman et al., 1998). Haemagglutinin and neuraminidase glycoproteins from influenza virus strains A and B are the antigens in this case. Inflexal V outperforms other conventional influenza vaccines in terms of immunogenicity and tolerability. Recently, liposomes have been shown to be beneficial.

have become more important in the battle against SARS-CoV-2. By safeguarding genetic material, lipid nanoparticles have effectively translated mRNA vaccines and ushered in a new era of gene therapy. Given the remarkable efficiency displayed by these vaccines, it is obvious that liposomal drug delivery is paving the way for future success in the delivery of genetic materials for a variety of therapeutic purposes. Similar to how the development of SARS-CoV-2 mRNA vaccines relied largely on optimising lipid composition, the development of these gene therapy-based drugs will as well (Large et al., 2021).

6.4. Liposomes in Gene Therapy and Diagnostics

Recombinant DNA technologies need the entry of nucleic acids into cells, both *in vitro* and *in vivo.* Gene function studies and gene therapy. In vivo delivery is more difficult than in vitro therapies, which may depend on a variety of physical and chemical methods. Numerous colloidal particles are present in DNA carrier systems: Negatively charged DNA is used in combination with cationic liposomes to transfect cells, allowing the target cells to produce the protein that is encoded in the DNA. In vivo delivery is preferred for gene therapy, which leads to the molecular-level treatment of diseases by turning on or off genes. It has been demonstrated that administering cationic lipid-based DNA complexes locally (mostly by intratracheal instillation of lung epithelial cells) or systemically (through lung endothelial cells) may transfect certain cells in vivo (Daree et al., 2016).

Audouy et al., developed cationic liposomes as delivery vectors for gene therapy in 2002 to investigate. Having a better understanding of the behaviour of lipoplexes and their ability to transfer genes in vivo paves the way for future therapeutic uses that will treat a wider range of conditions (Audouy et al., 2002). Liposomes act as a vehicle for viral glycolipids and glycoproteins, which are, co-adjuvants which enhance an immune response to the vaccination antigen. the development of liposomes is greatly hindered, of the liposomes made from commercially available lipids for the mucosal/oral administration of small molecules due to their vulnerability to enzymatic oxidation. The stomach's low pH and the intestine's dissolution of bile salts. As a result, common liposomes are unable to protect macromolecules from enzymatic breakdown of the digestive tract. Although they can tolerate these conditions, mechanically and sterically stabilised liposomes are too stable in the colon to release the drug. They are carried during regular absorption. Alternately, various capsules can be used to transport liposomes (or their precursors) directly to the colon. Given that it is generally known that discrete micelles, mixed micelles, and vesicles aid in the transportation of hydrophobic substances through the epithelial membranes, it is plausible that the mere presence of lipids increases absorption in the intestine. Lipids may also inhibit several enzymes necessary for the absorption or release of medications (Daree et al., 2016).

6.5. Liposomes in Cosmetics and Personal Care

Liposomes have been used for more than only drug delivery in the last 30 years; they are now the most widely used technique of delivering cosmetics. Liposomes can be used as a delivery mechanism because of their distinctive structure. They can transport hydrophilic medications (Laouini et al., 2012; Reva et al., 2015).

In cosmetics, an example is a fat-soluble stable molecule is vitamin E, used for its anti-aging, improved skin moisturising, and capacity to defend against skin illnesses (Ganesan et al.,2016). It is recommended to use a less water-soluble substance, like (LA) linoleic acid, to achieve whitening effects on hyperpigmented skin. Boosters for the skin-whitening effects of linoleic acid are liposomal preparations. The containing retinal acid or vitamin E liposomes, which may slow ascorbic acid oxidation, may be used in skin whitening compositions (Rieger et al., 2017). In terms of clinical treatment of acne vulgaris, liposomal lotions are more effective than non-liposomal lotions, particularly when treating pustules, where the clinical improvement was 75%. A conventional solution of lotion, an emulsion lotion without liposomes, and an emulsion lotion decreased the total number of lesions by 42.9, 48.3, and 62.8%, respectively, during a 4-week therapy period. The outcome points to the possibility of creating topical therapeutic solutions that are superior to those currently on the market using the liposomal dose form (Skalko et al., 1992). The development of innovative nanoparticulate systems to regulate the release to skin is the focus of current cosmetics research. Of all the molecules, liposomes are undoubtedly the most well-known. Since the structure and makeup of liposomes are very similar to those of the stratum corneum, administering this vehicle through the skin causes the lipidic components to be deposited, allowing the effects to last longer (Hua, 2015). When compared to goods including retarding agents, liposomes provide a slower release of the encapsulated content, making them an acceptable replacement for commercially available products. Due to its pleasant aroma and flavourings, limonene is frequently used in both food and cosmetic products (Sarisik et al., 2015).

UV radiation can have a variety of immediate and long-term effects on the skin. The initial reaction to UV light is erythema. Sun exposure over an extended period results in premature ageing and photocarcinogenesis, which are believed to be caused by genetic alterations and immune-mediated suppression (Nikolic et al., 2011). The amount of lipid present and the size of the particle both affect the UV-blocking abilities of lipid delivery systems like liposomes. With smaller particle size, the sunscreen action rises. Carnauba wax/decyloleate showed better and improved Skin Protection Factor (SPF) values in *in vitro* experiments when the investigated pigment Bismuith sulphate was incorporated into the lipid matrix of it (Muller et al., 2007).

When skin melanocytes are lost, the acquired idiopathic disorder known as vitiligo results, which is characterised by clearly defined milky white macules. It could significantly lower someone's quality of life, and in certain cases, it might even motivate suicide attempts (Nogueira et al., 2009). De Leeuw et al.,2011 studied the effect of liposomal khellin, a chemical identical to psoralen but with far less negative effects, on patient contentment following epidermal blister graft transplantation. Being enclosed in a liposome facilitates its entry into the hair. 76% of the patients were satisfied with the cosmetic result of Stimulating the melanocytes that occurs due to khellin present in the hair by UV light.

6.6. Liposomes in Food and Nutraceuticals

It is important to see how liposome technology is used in both practical and culinary applications. Liposomes were used to encapsulate clove oils to prevent chemical instability and enhance antibacterial action. Using liposome technology, clove essential oil stabilised and had a stronger antibacterial impact before being added to tofu. Staphylococcus aureus resistance was improved in tofu with better liposomes (Cui et al., 2015). Liposome technology has been used in milk and milk products for a long time. processing and enhancing dairy products using a variety of enzymes, including proteinases, lipases, nisin, and flavour-encapsulated liposome (Khanniri et al., 2016). Jahadi et al., 2015 used a flavoenzyme-containing nanoliposome in a dairy application. This is mostly used to block casein's early proteolysis and stop curd from maturing too quickly. Edible coating solutions based on hydroxypropyl methylcellulose have been developed using a novel approach for fatty foods such as chocolate and almonds. Vesicles were supplemented with rutin, a flavonoid having less molecular weight with strong antioxidant properties. By coating the thin surfaces of chocolate and almonds with this compound, their shelf life may be prolonged (Lopez-Polo et al., 2020). Sebaaly et al., 2021 investigated that extended release of drug from chitosomes had cutaneous penetration of drugs enhanced mucoadhesivity compared to uncoated liposomes. Chitosomes increased the biological effect and bioavailability. Curcumin has potent antioxidant effects, as well as various therapeutic qualities, including anti-inflammatory and antibacterial activities. Curcumin has been established to be encapsulated in a nanoliposome formulation to increase its stability and bioactivity. However, its bioactivities are constrained by its low water solubility, chemical instability, and photosensitivity. These limitations were overcome by making curcumin-containing pro-liposomes from micronised sucrose and non-purified phospholipids. Curcumin was abundant and there was less hygroscopicity in the proliposome powder (Liu et al., 2017). The many anthocyanin plant pigments give fruits and flowers their unique hues. Anthocyanins are known as BACs, and one anthocyanin form, cyanidin3-o-glucoside (C3G), has been extensively studied for its health benefits including its, anticancer potential, anti-diabetic, antioxidant, and anti-inflammatory (Rupasinghe et al., 2018). Therefore, it is clear that the use of liposome technology is a critical strategy to improve the stability, long-term release, and bioactivity of bioactive substances.

6.7. Liposomes in Biotechnology and Research

The idea of gene therapy requires reliable therapeutic gene delivery. In addition to aiding nucleic acid transport to the proper intracellular compartment and ensuring precise tissue targeting, the delivery technique should safeguard the nucleic acid against extracellular oxidation. In terms of protection, focussing on certain cells, and RNA delivery into cytoplasm, liposomal formulations can meet these criteria. Despite years of study and development, they tend to have drawbacks that reduce the efficiency of their medical applications. The presence of positively charged lipids in the liposome formulation strengthens the bond between the content (nucleic acids), which typically have a negative charge, and the liposomal carrier (Zuber et al., 2001).

In a variety of experiments to study how UV radiation affects skin, liposomes synthesised from linolenic acid, ceramide III or ceramide IV, DPPC and cholesterol were used as skin model membranes. These models were also used to study the redox behaviour of molecules like hyaluronan and ascorbic acid on skin when exposed to UV light. SCL liposomes have also been used as a model membrane to examine the mechanism of action of phospholipid vesicles as cutaneous drug delivery devices (Kirjavainen et al., 1996). Then, using dialysis using detergent techniques, and efforts were made to pack plasmids in liposomal devices with minuscule sizes and low surface charge when cationic lipid concentrations were low. These efforts drew on experience with the delivery of anti-cancer medicine in the form of a minute molecule (Wheeler et al., 1999). The discovery of hepatocyte gene silencing sparked intense research into the mechanism of action and the creation of stronger systems. These initiatives had remarkable success. For instance, demonstration shows the use of Potency increases of more than two orders of magnitude can be attained by using cationic lipids that are ionizable with a greater capacity to create non-bilayer structure and pKa values close to 6.6 (Semple et al., 2010). It has proven difficult to show the activity of nucleic acid in extrahepatic tissues. Extrahepatic targeting was initially shown using a formulation of long-circulating cationic liposomes (CCL) (Pagnan et al., 2000).

7. Challenges and Future Directions

Although there is an increasing number of nanotherapeutics in drug development pipelines, the effectiveness of clinical translation is not up to par. The ineffectiveness of clinical nanotherapeutic translation suggests that there are still issues that require attention (Zhang et al., 2020). Each nanotherapeutic has a unique set of difficulties during clinical translation, although these difficulties are universal to all nanotherapeutics: safety, biological difficulties (such as biodistribution), scale-up / cost, and regulation. Nanotherapeutic safety concerns are complicated. Clinical translation requires a thorough review of the safety of nanotherapeutics, but there are presently no established techniques or criteria for doing this. However, on the goods that are already on the market, we can state that liposomes have unquestionably cemented their place in contemporary technology. Although conventional medications continue to dominate the market, they are increasingly using nanotechnologies to lessen adverse effects and boost effectiveness. Liposomes were the first drug delivery device to receive clinical approval due to their biocompatibility and biodegradability. Despite years of study and development, they tend to have drawbacks that reduce the efficiency of their therapeutic uses. Improved payload, longer half-life, specialised transportation to the site of action, and the ability to circumvent chemotherapeutic resistance should all be advantages of liposomal-based medications. To develop liposomal-based treatments, preclinical animal models should analyse the differences between the current generation of liposomes and typical vesicles in terms of drug toxicity, tissue drug accumulation, and blood drug circulation timings. Additionally, it is important to emphasise the potentially dangerous side effects and clinical therapeutic advantages (Chang et al., 2012).

Conclusion

For an array of therapeutic payloads and applications in medicine, liposomes are the ideal carriers. because of the specific properties and functionality that are impacted by the liposome composition. Liposomes have been used for delivery for approximately 55 years after their development. Clinical uses of liposomes have significantly decreased off-target toxicity of numerous medications, allowed extended blood circulation, and improved drug biodistribution. Due to the benefits of nanotherapeutics in terms of pharmacokinetics, therapeutic effectiveness, and safety, the nano-pharmaceutical sector has experienced significant growth and development during the last 10 years. The huge potential and variety of the system must be explored for upcoming areas are demonstrated by a great number of methods available to produce liposomes. It is obvious that the research is heading towards more complicated liposome compositions and advanced targeting and release mechanisms. In this chapter, the analysis led to the conclusion that liposomes are highly effective as a drug delivery mechanism. Both hydrophilic and lipophilic drugs are easily encapsulated in liposomes. The medicine was administered to the patient in a regulated manner or was targeted at a certain place. By using liposomes, the medication was administered efficiently and effectively in several difficult disorders (cancers, tumours, and HIV).

References

Abra RM, Bosworth ME, Hunt CA. Liposome disposition *in vivo*: effects of pre-dosing with liposomes. *Res Commun Chem Pathol Pharmacol.* 1980;29(2):349-360.

Akashi KI, Miyata H, Itoh H, Kinosita K. Preparation of giant liposomes in physiological conditions and their characterization under an optical microscope. *Biophys J.* 1996;71(6):3242-3250.

Akbarzadeh A, Rezaei-Sadabady R, Davaran S, Joo SW, Zarghami N, Hanifehpour Y, Samiei M, Kouhi M, Nejati-Koshki K. Liposome: classification, preparation, and applications. *Nanoscale Res Lett.* 2013;8:1-9.

Allen TM, Chonn A. Large unilamellar liposomes with low uptake into the reticuloendothelial system. *FEBS Lett.* 1987;223(1):42-46.

Audouy SA, de Leij LF, Hoekstra D, Molema G. In vivo characteristics of cationic liposomes as delivery vectors for gene therapy. *Pharm Res.* 2002;19:1599-1605.

Bangham AD, Standish MM, Watkins JC. Diffusion of univalent ions across the lamellae of swollen phospholipids. *J Mol Biol.* 1965;13(1):238-IN27.

Batzri S, Korn ED. Single bilayer liposomes prepared without sonication. *Biochim Biophys Acta Biomembr*. 1973;298(4):1015-1019.

Biner O, Schick T, Müller Y, von Ballmoos C. Delivery of membrane proteins into small and giant unilamellar vesicles by charge-mediated fusion. *FEBS Lett.* 2016;590(14):2051-2062.

Bulbake U, Doppalapudi S, Kommineni N, Khan W. Liposomal formulations in clinical use: an updated review. *Pharmaceutics*. 2017;9(2):12.

Chang HI, Yeh MK. Clinical development of liposome-based drugs: formulation, characterization, and therapeutic efficacy. *Int J Nanomedicine*. 2012;7:49-60.

Charcosset C, Juban A, Valour JP, Urbaniak S, Fessi H. Preparation of liposomes at large scale using the ethanol injection method: Effect of scale-up and injection devices. *Chem Eng Res Des*. 2015;94:508-515.

Chatterjee SN, Devhare PB, Lole KS. Detection of negative-sense RNA in packaged hepatitis E virions by use of an improved strand-specific reverse transcription-PCR method. *J Clin Microbiol*. 2012;50(4):1467-1470.

Chaves MA, Oseliero Filho PL, Jange CG, Sinigaglia-Coimbra R, Oliveira CLP, Pinho SC. Structural characterization of multilamellar liposomes coencapsulating curcumin and vitamin D3. *Colloids Surf A Physicochem Eng Asp*. 2018;549:112-121.

Chen H, Gao J, Wang F, Liang W. Preparation, characterization and pharmacokinetics of liposomes-encapsulated cyclodextrins inclusion complexes for hydrophobic drugs. *Drug Deliv*. 2007;14(4):201-20.

Chen J, Lu WL, Gu W, Lu SS, Chen ZP, Cai BC. Skin permeation behavior of elastic liposomes: role of formulation ingredients. *Expert Opin Drug Deliv*. 2013;10(6):845-856.

Chen SH, Liu XW, Zhan SP, Yu C, Zhang J. Preparation and characterization of metronidazole liposome by supercritical reverse phase evaporation method. *Adv Mater Res*. 2013;668:274-278.

Chorilli M, Calixto G, Rimerio TC, Scarpa MV. Caffeine encapsulated in small unilamellar liposomes: characterization and in vitro release profile. *J Dispersion Sci Technol.* 2013;34(10):1465-1470.

Clark PR, Stopeck AT, Ferrari M, Parker SE, Hersh EM. Studies of direct intratumoral gene transfer using cationic lipid-complexed plasmid DNA. *Cancer Gene Ther.* 2000;7(6):853-860.

Cosco D, Bulotta A, Ventura M, Celia C, Calimeri T, Perri G, Paolino D, Costa N, Neri P, Tagliaferri P, Tassone P. In vivo activity of gemcitabine-loaded PEGylated small unilamellar liposomes against pancreatic cancer. *Cancer Chemother Pharmacol.* 2009;64:1009-1020.

Cui H, Zhao C, Lin L. The specific antibacterial activity of liposome-encapsulated Clove oil and its application in tofu. *Food Control*. 2015;56:128-134.

Čuperlović-Culf M. Biology—Cancer Metabolic Phenotype. In: *NMR Metabolomics in Cancer Research*. 2013. p. 15-138.

Daemen T, De Haan A, Arkema A, Wilschut J. Liposomes and virosomes as immunoadjuvant and antigen-carrier systems in vaccine formulations. *Med Appl Liposomes*. 1998. p. 117-143.

Daraee H, Etemadi A, Kouhi M, Alimirzalu S, Akbarzadeh A. Application of liposomes in medicine and drug delivery. *Artif Cells Nanomed Biotechnol*. 2016;44(1):381-391.

De Leeuw J, Assen YJ, Van Der Beek N, Bjerring P, Martino Neumann HA. Treatment of vitiligo with khellin liposomes, ultraviolet light and blister roof transplantation. *J Eur Acad Dermatol Venereol*. 2011;25(1):74-81.

De Leo V, Ruscigno S, Trapani A, Di Gioia S, Milano F, Mandracchia D, Comparelli R, Castellani S, Agostiano A, Trapani G, Catucci L. Preparation of drug-loaded small unilamellar liposomes and evaluation of their potential for the treatment of chronic respiratory diseases. *Int J Pharm*. 2018;545(1-2):378-388.

Deamer D, Bangham AD. Large volume liposomes by an ether vaporization method. Biochim Biophys Acta. 1976;443(3):629-634.

Deamer DW. From "banghasomes" to liposomes: a memoir of Alec Bangham, 1921-2010. *FASEB J*. 2010;24(5):1308.

Detoni CB, Cabral-Albuquerque ECM, Hohlemweger SVA, Sampaio C, Barros TF, Velozo ES. Essential oil from Zanthoxylum tingoassuiba loaded into multilamellar liposomes useful as antimicrobial agents. *J Microencapsulation*. 2009;26(8):684-691.

Dorrani M, Garbuzenko OB, Minko T, Michniak-Kohn B. Development of edge-activated liposomes for siRNA delivery to human basal epidermis for melanoma therapy. *J Control Release*. 2016;228:150-158.

Dua JS, Rana AC, Bhandari AK. Liposome: methods of preparation and applications. *Int J Pharm Stud Res*. 2012;3(2):14-20.

El Maghraby GMM, Williams AC, Barry BW. Interactions of surfactants (edge activators) and skin penetration enhancers with liposomes. *Int J Pharm*. 2004;276(1-2):143-161.

El Zaafarany GM, Awad GA, Holayel SM, Mortada ND. Role of edge activators and surface charge in developing ultradeformable vesicles with enhanced skin delivery. *Int J Pharm*. 2010;397(1-2):164-172.

Elsana H, Olusanya TO, Carr-Wilkinson J, Darby S, Faheem A, Elkordy AA. Evaluation of novel cationic gene-based liposomes with cyclodextrin prepared by thin film hydration and microfluidic systems. *Sci Rep*. 2019;9(1):15120.

Farooque F, Wasi M, Mughees MM. Liposomes as Drug Delivery System: An Updated Review. *J Drug Deliv Ther*. 2021;11(5-S):149-158.

Fisher LR, Mitchell EE, Parker NS. Interfacial tensions of commercial vegetable oils with water. *J Food Sci*. 1985;50(4):1201-1202.

Forssen EA, Tökès ZA. Use of anionic liposomes for the reduction of chronic doxorubicin-induced cardiotoxicity. *Proc Natl Acad Sci*. 1981;78(3):1873-1877.

Ganesan P, Choi DK. Current application of phytocompound-based nanocosmeceuticals for beauty and skin therapy. *Int J Nanomedicine*. 2016;11:1987-2007.

Glück R, Mischler R, Finkel B, Que JU, Cryz Jr SJ, Scarpa BJTL. Immunogenicity of new virosome influenza vaccine in elderly people. *Lancet*. 1994;344(8916):160-163.

Gorter F, Grendel FJEM. On bimolecular layers of lipoids on the chromocytes of the blood. *J Exp Med*. 1925;41(4):439.

Gregoriadis G, Perrie Y. *Liposomes*. eLS. 2010.

Gregoriadis G, Ryman BE. Fate of protein-containing liposomes injected into rats: An approach to the treatment of storage diseases. *Eur J Biochem.* 1972;24(3):485-491.

Gregoriadis G, Ryman BE. Lysosomal localization of β-fructofuranosidase-containing liposomes injected into rats. Some implications in the treatment of genetic disorders. *Biochem J.* 1972;129(1):123-133.

Gregoriadis G. *Liposome technology*. CRC Press: Boca Raton, FL, USA. 1992; Vols. I–III.

Gregoriadis G. Liposomes in drug delivery: how it all happened. *Pharmaceutics.* 2016;8(2):19.

Group1A VIPTS. Verteporfin therapy of subfoveal choroidal neovascularization in age-related macular degeneration: two-year results of a randomized clinical trial including lesions with occult with no classic choroidal neovascularization—verteporfin in photodynamic therapy report 2. *Am J Ophthalmol.* 2001;131(5):541-560.

Guo P, You JO, Yang J, Jia D, Moses MA, Auguste DT. Inhibiting metastatic breast cancer cell migration via the synergy of targeted, pH-triggered siRNA delivery and chemokine axis blockade. *Mol Pharm.* 2014;11(3):755-765.

Haran G, Cohen R, Bar LK, Barenholz Y. Transmembrane ammonium sulfate gradients in liposomes produce efficient and stable entrapment of amphipathic weak bases. *Biochim Biophys Acta (BBA)-Biomembranes.* 1993;1151(2):201-215.

He H, Yuan D, Wu Y, Cao Y. Pharmacokinetics and pharmacodynamics modeling and simulation systems to support the development and regulation of liposomal drugs. *Pharmaceutics*. 2019;11(3):110.

Heath TD, Montgomery JA, Piper JR, Papahadjopoulos D. Antibody-targeted liposomes: increase in specific toxicity of methotrexate-gamma-aspartate. *Proc Natl Acad Sci.* 1983;80(5):1377-1381.

Hofheinz RD, Gnad-Vogt SU, Beyer U, Hochhaus A. Liposomal encapsulated anti-cancer drugs. *Anticancer Drugs*. 2005;16(7):691-707.

Hua S. Lipid-based nano-delivery systems for skin delivery of drugs and bioactives. *Front Pharmacol*. 2015;6:219.

Hunter DG, Frisken BJ. Effect of extrusion pressure and lipid properties on the size and polydispersity of lipid vesicles. *Biophys J.* 1998;74(6):2996-3002.

Immordino ML, Dosio F, Cattel L. Stealth liposomes: review of the basic science, rationale, and clinical applications, existing and potential. *Int J Nanomedicine.* 2006;1(3):297.

Imura T, Otake K, Hashimoto S, Gotoh T, Yuasa M, Yokoyama S, Sakai H, Rathman JF, Abe M. Preparation and physicochemical properties of various soybean lecithin liposomes using supercritical reverse phase evaporation method. *Colloids Surf B Biointerfaces*. 2003;27(2-3):133-140.

Jaafar-Maalej C, Diab R, Andrieu V, Elaissari A, Fessi H. Ethanol injection method for hydrophilic and lipophilic drug-loaded liposome preparation. *J Liposome Res.* 2010;20(3):228-243.

Jahadi M, Khosravi-Darani K, Ehsani MR, Mozafari MR, Saboury AA, Pourhosseini PS. The encapsulation of flavourzyme in nanoliposome by heating method. *J Food Sci Technol*. 2015;52:2063-2072.

Johnston MJ, Semple SC, Klimuk SK, Ansell S, Maurer N, Cullis PR. Characterization of the drug retention and pharmacokinetic properties of liposomal nanoparticles containing dihydrosphingomyelin. *Biochim Biophys Acta (BBA)-Biomembranes.* 2007;1768(5):1121-1127.

Joo KI, Xiao L, Liu S, Liu Y, Lee CL, Conti PS, Wong MK, Li Z, Wang P. Crosslinked multilamellar liposomes for controlled delivery of anticancer drugs. *Biomaterials.* 2013;34(12):3098-3109.

Justo OR, Moraes ÂM. Analysis of process parameters on the characteristics of liposomes prepared by ethanol injection with a view to process scale-up: Effect of temperature and batch volume. *Chem Eng Res Des.* 2011;89(6):785-792.

Kataria S, Sandhu P, Bilandi AJAY, Akanksha, Kapoor B. Stealth liposomes: a review. *Int J Res Ayurveda Pharm.* 2011;2(5).

Khanniri E, Bagheripoor-Fallah N, Sohrabvandi S, Mortazavian AM, Khosravi-Darani K, Mohammad R. Application of liposomes in some dairy products. *Crit Rev Food Sci Nutr.* 2016;56(3):484-493.

Kirby C, Clarke J, Gregoriadis G. Effect of the cholesterol content of small unilamellar liposomes on their stability in vivo and in vitro. *Biochem J.* 1980;186(2):591-598.

Kirjavainen M, Urtti A, Jääskeläinen I, Suhonen TM, Paronen P, Valjakka-Koskela R, Kiesvaara J, Mönkkönen J. Interaction of liposomes with human skin in vitro—the influence of lipid composition and structure. *Biochim Biophys Acta (BBA)-Lipids Lipid Metab.* 1996;1304(3):179-189.

Klibanov AL, Maruyama K, Beckerleg AM, Torchilin VP, Huang L. Activity of amphipathic poly (ethylene glycol) 5000 to prolong the circulation time of liposomes depends on the liposome size and is unfavorable for immunoliposome binding to target. *Biochim Biophys Acta (BBA)-Biomembranes.* 1991;1062(2):142-148.

Laouini A, Jaafar-Maalej C, Limayem-Blouza I, Sfar S, Charcosset C, Fessi H. Preparation, characterization and applications of liposomes: state of the art. *J Colloid Sci Biotechnol.* 2012;1(2):147-168.

Large DE, Abdelmessih RG, Fink EA, Auguste DT. Liposome composition in drug delivery design, synthesis, characterization, and clinical application. *Adv Drug Deliv Rev.* 2021;176:113851.

Leserman L. The Gregoriadyssey: from smetic mesophases to liposomal drug carriers, a personal reflection of Gregory Gregoriadis. *J Drug Target.* 2008;16(7-8):525-528.

Liu G, Hou S, Tong P, Li J. Liposomes: preparation, characteristics, and application strategies in analytical chemistry. *Crit Rev Anal Chem.* 2022;52(2):392-412.

Liu Y, Ying D, Cai Y, Le X. Improved antioxidant activity and physicochemical properties of curcumin by adding ovalbumin and its structural characterization. *Food Hydrocolloids.* 2017;72:304-311.

Lopez-Polo J, Silva-Weiss A, Zamorano M, Osorio FA. Humectability and physical properties of hydroxypropyl methylcellulose coatings with liposome-cellulose nanofibers: Food application. *Carbohydr Polym.* 2020;231:115702.

Mayer LD, Tai LC, Bally MB, Mitilencs GN, Ginsberg RS, Cullis PR. Characterization of liposomal systems containing doxorubicin entrapped in response to pH gradients. *Biochim Biophys Acta (BBA)-Biomembranes.* 1990;1025(2):143-151.

Mc Cauley JA. Flory's Book.; Mc Comb TG. *Biochim. Biophys. Acta.* 1992;30:112.

Mischler R, Metcalfe IC. Inflexal® V a trivalent virosome subunit influenza vaccine: production. *Vaccine*. 2002;20:B17-B23.

Mori A, Klibanov AL, Torchilin VP, Huang L. Influence of the steric barrier activity of amphipathic poly (ethyleneglycol) and ganglioside GM1 on the circulation time of liposomes and on the target binding of immunoliposomes in vivo. *FEBS Lett.* 1991;284(2):263-266.

Müller R, Souto E, Alemieda AJ. Topical delivery of oily actives using solid lipid particles. *Pharm Technol Eur*. 2007;19(12).

New RRC, Chance ML, Thomas SC, Peters W. Antileishmanial activity of antimonials entrapped in liposomes. *Nature*. 1978;272(5648):55-56.

Nikolić S, Keck CM, Anselmi C, Müller RH. Skin photoprotection improvement: synergistic interaction between lipid nanoparticles and organic UV filters. *Int J Pharm*. 2011;414(1-2):276-284.

Nogueira LS, Zancanaro PC, Azambuja RD. Vitiligo e emoções. *An Bras Dermatol.* 2009;84:41-45.

Okamoto Y, Taguchi K, Yamasaki K, Sakuragi M, Kuroda SI, Otagiri M. Albumin-encapsulated liposomes: a novel drug delivery carrier with hydrophobic drugs encapsulated in the inner aqueous core. *J Pharm Sci*. 2018;107(1):436-445.

Olson F, Hunt CA, Szoka FC, Vail WJ, Papahadjopoulos D. Preparation of liposomes of defined size distribution by extrusion through polycarbonate membranes. *Biochim Biophys Acta (BBA)-Biomembranes*. 1979;557(1):9-23.

Pagnan G, Stuart DD, Pastorino F, Raffaghello L, Montaldo PG, Allen TM, Calabretta B, Ponzoni M. Delivery of c-myb antisense oligodeoxynucleotides to human neuroblastoma cells via disialoganglioside GD2-targeted immunoliposomes: antitumor effects. *Journal of the National Cancer Institute*. 2000 Feb 2;92(3):253-61.

Papahadjopoulos D, Gabizon A. Targeting of Liposomes to Tumor Cells in Vivo a. *Ann N Y Acad Sci*. 1987;507(1):64-74.

Patil SD, Rhodes DG, Burgess DJ. Anionic liposomal delivery system for DNA transfection. *AAPS J.* 2004;6:13-22.

Perez AP, Altube MJ, Schilrreff P, Apezteguia G, Celes FS, Zacchino S, de Oliveira CI, Romero EL, Morilla MJ. Topical amphotericin B in ultradeformable liposomes: Formulation, skin penetration study, antifungal and antileishmanial activity *in vitro*. *Colloids and Surfaces B: Biointerfaces*. 2016 Mar 1;139:190-8.

Perrett S, Golding M, Williams WP. A simple method for the preparation of liposomes for pharmaceutical applications: characterization of the liposomes. *J Pharm Pharmacol*. 1991;43(3):154-161.

Pham HL, Shaw PN, Davies NM. Preparation of immuno-stimulating complexes (ISCOMs) by ether injection. *Int J Pharm*. 2006;310(1-2):196-202.

Pons M, Foradada M, Estelrich J. Liposomes obtained by the ethanol injection method. *Int J Pharm.* 1993;95(1-3):51-56.

Reva T, Vaseem A, Satyaprakash S, Khalid JM. Liposomes: the novel approach in cosmaceuticals. *World J Pharm Pharm Sci*. 2015;4(6):1616-1640.

Riaz M. Liposomes preparation methods. *Pak J Pharm Sci*. 1996;9(1):65-77.

Rieger M, ed. *Surfactants in cosmetics*. Routledge; 2017.

Rigaud JL, Bluzat A, Buschlen S. Incorporation of bacteriorhodopsin into large unilamellar liposomes by reverse phase evaporation. *Biochem Biophys Res Commun.* 1983;111(2):373-382.

Rodrigues C, Gameiro P, Prieto M, De Castro B. Interaction of rifampicin and isoniazid with large unilamellar liposomes: spectroscopic location studies. *Biochim Biophys Acta (BBA)-General Subjects.* 2003;1620(1-3):151-159.

Rupasinghe HV, Arumuggam N, Amararathna M, De Silva ABKH. The potential health benefits of haskap (Lonicera caerulea L.): Role of cyanidin-3-O-glucoside. *J Funct Foods.* 2018;44:24-39.

Safari J, Zarnegar Z. Advanced drug delivery systems: Nanotechnology of health design A review. *J Saudi Chem Soc.* 2014;18(2):85-99.

SARIIŞIK A, KARTAL G. Disposable mask design for odor pollution in the work environment. *Tekstil ve Mühendis.* 2015;22(97):30-36.

Schieren H, Rudolph S, Finkelstein M, Coleman P, Weissmann G. Comparison of large unilamellar vesicles prepared by a petroleum ether vaporization method with multilamellar vesicles: ESR, diffusion and entrapment analyses. *Biochim Biophys Acta (BBA)-General Subjects.* 1978;542(1):137-153.

Schroeder U, Sommerfeld P, Ulrich S, Sabel BA. Nanoparticle technology for delivery of drugs across the blood–brain barrier. *J Pharm Sci.* 1998;87(11):1305-1307.

Sebaaly C, Trifan A, Sieniawska E, Greige-Gerges H. Chitosan-coating effect on the characteristics of liposomes: A focus on bioactive compounds and essential oils: A review. *Processes.* 2021;9(3):445.

Semple SC, Akinc A, Chen J, Sandhu AP, Mui BL, Cho CK, Sah DW, Stebbing D, Crosley EJ, Yaworski E, Hafez IM. Rational design of cationic lipids for siRNA delivery. *Nature biotechnology.* 2010 Feb;28(2):172-6.

Sessa G, Weissmann G. Incorporation of lysozyme into liposomes: a model for structure-linked latency. *J Biol Chem.* 1970;245(13):3295-3301.

Singer SJ, Nicolson GL. The Fluid Mosaic Model of the Structure of Cell Membranes: Cell membranes are viewed as two-dimensional solutions of oriented globular proteins and lipids. *Science.* 1972;175(4023):720-731.

Škalko N, Čajkovac M, Jalšenjak I. Liposomes with clindamycin hydrochloride in the therapy of acne vulgaris. *Int J Pharm.* 1992;85(1-3):97-101.

Suri SS, Fenniri H, Singh B. Nanotechnology-based drug delivery systems. *J Occup Med Toxicol.* 2007;2:1-6.

Tagrida M, Prodpran T, Zhang B, Aluko RE, Benjakul S. Liposomes loaded with betel leaf (Piper betle L.) ethanolic extract prepared by thin film hydration and ethanol injection methods: Characteristics and antioxidant activities. *J Food Biochem.* 2021;45(12):e14012.

Tazina EV, Kostin KV, Oborotova NA. Specific features of drug encapsulation in liposomes (A review). *Pharm Chem J.* 2011;45(8):481-490.

Tokudome Y, Saito Y, Sato F, Kikuchi M, Hinokitani T, Goto K. Preparation and characterization of ceramide-based liposomes with high fusion activity and high membrane fluidity. *Colloids Surf B Biointerfaces.* 2009;73(1):92-96.

Torchilin VP, Weissig V, eds. *Liposomes: a practical approach* (No. 264). Oxford University Press; 2003.

Tsumoto K, Matsuo H, Tomita M, Yoshimura T. Efficient formation of giant liposomes through the gentle hydration of phosphatidylcholine films doped with sugar. *Colloids Surf B Biointerfaces*. 2009;68(1):98-105.

Wang T, Deng Y, Geng Y, Gao Z, Zou J, Wang Z. Preparation of submicron unilamellar liposomes by freeze-drying double emulsions. *Biochim Biophys Acta (BBA)-Biomembr*. 2006;1758(2):222-231.

Watson DS, Endsley AN, Huang L. Design considerations for liposomal vaccines: influence of formulation parameters on antibody and cell-mediated immune responses to liposome-associated antigens. *Vaccine*. 2012;30(13):2256-2272.

Weiner AL. Lamellar systems for drug solubilization. *Liposomes: from biophysics to therapeutics.* Marcel Dekker, New York. 1987:339-69.

Wheeler JJ, Palmer L, Ossanlou M, MacLachlan I, Graham RW, Zhang YP, Hope MJ, Scherrer P, Cullis PR. Stabilized plasmid-lipid particles: construction and characterization. *Gene therapy*. 1999 Feb;6(2):271-81.

Yadav AMMS, Murthy MS, Shete AS, Sakhare S. Stability aspects of liposomes. *Indian J Pharm Educ Res.* 2011;45(4):402-413.

Yang G, Yang T, Zhang W, Lu M, Ma X, Xiang G. In vitro and in vivo antitumor effects of folate-targeted ursolic acid stealth liposome. *J Agric Food Chem.* 2014;62(10): 2207-2215.

Zeb A, Qureshi OS, Kim HS, Cha JH, Kim HS, Kim JK. Improved skin permeation of methotrexate via nanosized ultradeformable liposomes. *Int J Nanomedicine.* 2016;11:3813-3824.

Zhang C, Yan L, Wang X, Zhu S, Chen C, Gu Z, Zhao Y. Progress, challenges, and future of nanomedicine. *Nano Today*. 2020;35:101008.

Zuber G, Dauty E, Nothisen M, Belguise P, Behr JP. Towards synthetic viruses. *Adv Drug Deliv Rev*. 2001;52(3):245-253.

Chapter 2

Liposome-Mediated Drug Delivery: Recent Developments and Challenges

Widhilika Singh[1]
and Poonam Kushwaha[1,*]
[1]Faculty of Pharmacy, Integral University, Lucknow, Uttar Pradesh, India

Abstract

Liposome-mediated therapeutics delivery is a versatile and promising scientific technology that favours the delivery of therapeutic agents to the target organ in a controlled fashion. The latest advances in liposome technology have led to improved drug charging and release, improved biocompatibility, and increased therapeutic efficacy of liposome formulations. However, challenges associated with liposome-mediated drug delivery still exist, including low drug encapsulation efficiency, limited drug stability, and suboptimal biodistribution. This chapter presents a comprehensive outline of the latest advances in the field of liposome technology and the associated difficulties that arise with drug delivery through liposomes. The discourse concerns the various determinants that impact the loading and dispensation of drugs from liposomes, along with the tactics to surmount the difficulties involved in the delivery of drugs through liposomes. Finally, this section of the book delves into future possibilities of drug delivery through liposomes and highlights the imperative requirement for persistent research in this domain to overcome obstacles and actualise the potential of this technology in treating various diseases.

[*] Corresponding Author's Email: poonam@iul.ac.in.

In: Liposomes
Editors: Usama Ahmad and Anas Islam
ISBN: 979-8-89113-636-6

Keywords: liposomes, drug delivery, nanoliposomes, drug loading, liposome challenges, liposome development

1. Introduction

Liposome-mediated drug administration is considered a propitious method for the transfer of therapeutic agents. This mechanism employs lipid bilayer vesicles, commonly known as liposomes, to envelop drugs and transport them to designated locations within the human body. Liposomes have been conduits for drug transfer since the 1970s. The past of liposome-mediated drug delivery is rich in developments and breakthroughs that have led to improved drug delivery systems (Table 1).

Liposomes refer to drug-containing vesicles produced by the self-arrangement of phospholipids, which have either a single bilayer or multiple concentric bilayers encompassing an inner aqueous core. The dimensions range from 30 nanometres to the micrometre scale and the phospholipid bilayer has a thickness of approximately 4-5 nanometres (Liu et al., 2022).

The vesicles that Bangham and his associates named liposomes during the 1960s could encapsulate both hydrophilic and hydrophobic molecules. The liposome technology remained in the laboratory for several years before it was used for drug delivery (Bangham and Horne, 1964; Bangham et al., 1965).

1.1. Advantages of Liposome-Mediated Drug Delivery Over Conventional and Other Pharmaceutical Delivery Systems

Drug transportation through the liposome is a versatile approach to administering therapeutic agents. Liposomes enclose drugs and deliver them to the aiming site in the biological system. Compared to conventional drug delivery systems, liposome-mediated drug delivery offers several advantages, including improved pharmacokinetics, reduced toxicity, and improved therapeutic efficacy (Table 2).

Table 1. List of liposome-based products with their year of approval

S. No.	Marketed Product	Active Pharmaceutical Ingredient	Formulation	Indication	Company Name	Year of Approval
1	Epaxal®	synthetic lipids, viral proteins, Hepatitis A immunogens	Hepatitis A Liposomal vaccine/ Virosome	Hepatitis	Crucell, Berna Biotech	1994 (2003-Europe, Asia)
2	Doxil®	Doxorubicin	Liposomal IV Injection/ Infusion	Ovarian cancer, HIV-associated Kaposi's sarcoma	Baxter Healthcare Corporation	1995
3	DaunoXome®a	Daunorubicin	Liposomal IV Injection/ Infusion	HIV-associated Kaposi's sarcoma	NeXstar Pharmaceuticals USA, Galen US Inc.	1996
4	Amphotec®	Amphotericin B	Lyophilised powder for reconstitution and intravenous (IV) administration.	Fungal infections	InterMune, Inc.	1996
5	Ambisome®	Amphotericin B	Liposome for injection	Fungal infections	Gilead Sciences, Astellas Pharma Inc.	1997
6	Inflexal®	Flu vaccine	Liposomal vaccine	Influenza	Crucell, Berna Biotech	1997
7	Depocyt®	Cytarabine	Liposomal IV Injection/ Infusion	Leukemia	SkyPharma Inc	1999
8	Myocet®	Doxorubicin	Liposomal IV Injection/ Infusion	Metastatic breast cancer	Elan Pharmaceuticals	2000
9	Visudyne®	Verteporfin	Injection	Photodynamic therapy / antineovascularisation agent for classic subfoveal choroidal neovascularization	Bausch and Lomb	2000
10	DepoDurTM	Morphine	Extended-Release Liposome Injection)	Analgesic	Skye Pharma Inc	2004
11	Mepact®	Mifamurtide	Liposomal IV Injection/ Infusion	Cancer (Osteosarcoma)	Takeda Pharmaceutica	2009
12	Exparel®	Bupivacaine	Liposome injectable suspension	Analgesic	Pacira	2011
13	Marqibo®	Vincristine	Liposomal IV Injection/ Infusion	Cancer	Talon Therapeutics	2012
14	Lipodox®	Doxorubicin	Liposome for injection	Cancer	Sun Pharmaceutical Industries Ltd.	2012

Table 1. (Continued)

S. No.	Marketed Product	Active Pharmaceutical Ingredient	Formulation	Indication	Company Name	Year of Approval
15	Onivyde™	Irinotecan	Liposomal IV Injection/ Infusion	Metastatic adenocarcinoma of the pancreas	Merrimack Pharmaceuticals, Inc	2015
16	Vyxeos®	Daunorubicin-cytarabine (1:5)	Liposome for injection	Cancer (Acute Myeloid Leukaemia)	Jazz Pharmaceuticals, Inc	2017
17	Arikayce®	Amikacin	Liposome Inhalational Suspension	Mycobacterium avium complex (MAC) lung disease	Insmed	2018
18	Onpattro®	Patisiran	Liposome IV Injection/ Infusion	Polyneuropathy associated with Transthyretin-Mediated (ATTR) amyloidosis	Alnylam Pharmaceuticals Inc	2018
19	COVID Vaccine	mRNA 1227 in Lipid Nanoparticle	Intramuscular Injection	SARS-CoV-19	Moderna	2020
20	COVID Vaccine	mRNA 1227 in Lipid Nanoparticle	Intramuscular Injection	SARS-CoV-19	Pfizer	2020

Table 2. Advantages of liposome-mediated drug delivery

S. No.	Aspect	Advantages of the Liposomal Drug Delivery System	Reference
1	*Improved Pharmacokinetics*	Improved Pharmacokinetics of the encapsulated drug Protection against degradation by enzymes and other physiological factors, such as pH and temperature, and physiological factors	(Torchilin, 2005)
		Altered distribution and clearance profiles	(Sawant and Torchilin, 2012)
		Prolonged presence in the bloodstream	(Allen and Cullis, 2013)
		Reduced uptake by the phagocytic system	(Sercombe et al., 2015)
		Stimuli-sensitive release and targeted delivery	(Sawant and Torchilin, 2012)
2	*Reduced Toxicity*	Reduction in drug toxicity	(Torchilin, 2005; Tiwari et al., 2012)
		Liposomal aerosol formulation of 9-Nitrocamptothecin as non-toxic in animal study	(Gilbert et al., 2002)
		Selective accumulation in target tissues, minimising drug exposure to off-target tissues	(Allen and Cullis, 2013; Tiwari et al., 2012)
		Targeting moieties on modified liposomes; antibodies or peptides that bind on particular target cell binding sites	(Torchilin, 2005)
		Reduced cardiac toxicity with liposomal doxorubicin administration	(Crommelin et al., 2020)
3	*Enhanced Therapeutic Efficacy*	Improved drug solubility and cellular uptake	(Teixeira et al., 2017)
		Overcoming solubility issues with liposomal formulations, as in case of voriconazole	(Veloso et al., 2018)
		Enhanced drug release profiles, sustained drug release, and prolonged therapeutic effects	(Torchilin, 2005; Allen and Cullis 2013; Maritim et al., 2021)
		Decreased need for frequent administration and improved adherence to treatment protocols	(Maritim et al., 2021)
		Enhanced delivery to tumour sites and extended release, as in case of daunorubicin	(Liu et al., 2022)
		Improved permeability across biological barriers, such as the blood-brain barrier (BBB)	(Agrawal et al., 2017)
4	*Delivery of radiotherapeutics*	Favourable pharmacokinetic profile with rapid blood clearance and low systemic exposure	(Wang et al., 2019)
		Selective accumulation of the 188Re-liposome in tumour lesions, indicating its potential as a targeted therapeutic agent	
		Dosimetry calculations demonstrated acceptable radiation dose delivery to tumor sites	
		Feasibility and safety of the 188Re-liposome for localized radiation therapy	

Table 2. (Continued)

S. No.	Aspect	Advantages of the Liposomal Drug Delivery System	Reference
		Potential clinical application of the 188Re-liposome as a targeted treatment option for metastatic tumors	
5	*Comparison to Other Drug Delivery Systems*	Highly biocompatible and biodegradable structures	(Torchilin, 2005; Almeida et al., 2020)
		Safety and well-tolerated drug delivery system	(Torchilin, 2005; Almeida et al., 2020)
		Absence of unexpected toxic or immune reactions	(Torchilin, 2005)
		Protected from the immune system by polyethylene glycol coating	(Zolnik et al., 2010)
		Precise control over physicochemical properties	(Torchilin, 2005; Nsairat et al., 2022)
		Modification of Properties for targeting approaches	(Torchilin, 2005; Nsairat et al., 2022)
		Optimal drug encapsulation, release profiles, targeting, and cellular uptake	(Torchilin, 2005; Düzgüneş and Nir, 1999)

2. Liposome composition and design considerations

The composition and design are significant factors in determining their performance, efficacy, and successful drug delivery system.

2.1. Liposome Composition

Lipids, steroids/ cholesterols, and other additives such as surfactants are used to formulate liposomes (Figure 1) (Nsairat et al., 2022).

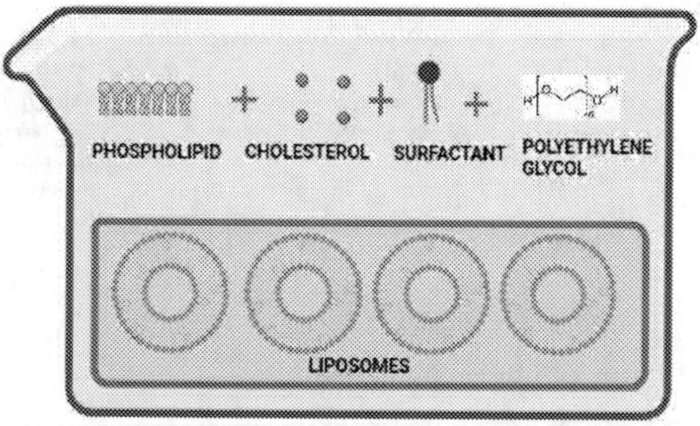

Figure 1. Liposome Composition.

2.1.1. Lipid Components

The selection of lipid components is an essential factor in the determination of the characteristics of liposomes (Torchilin, 2005; Patel and Sprott, 1999). Phospholipids, such as phosphatidylcholine and phosphatidylglycerol, are frequently used to formulate a liposome that provides structural integrity to the liposome (Li et al., 2015). The selection of specific phospholipids can affect the stability of the liposome, the efficiency of drug encapsulation, and release kinetics (Figure 2) (Barenholz, 2001; Anderson and Omri, 2004).

2.1.2. Cholesterol

The addition of cholesterol to liposomes is essential to modulate their physical properties (Akbarzadeh et al., 2013). Cholesterol improves the membrane rigidity and stability of the membrane of liposomes, while also influencing their fluidity and permeability (Kaddah et al., 2018; Nsairat et al., 2022). The optimal cholesterol concentration must be carefully determined to achieve the desired characteristics of the liposomes (Shaker et al., 2017).

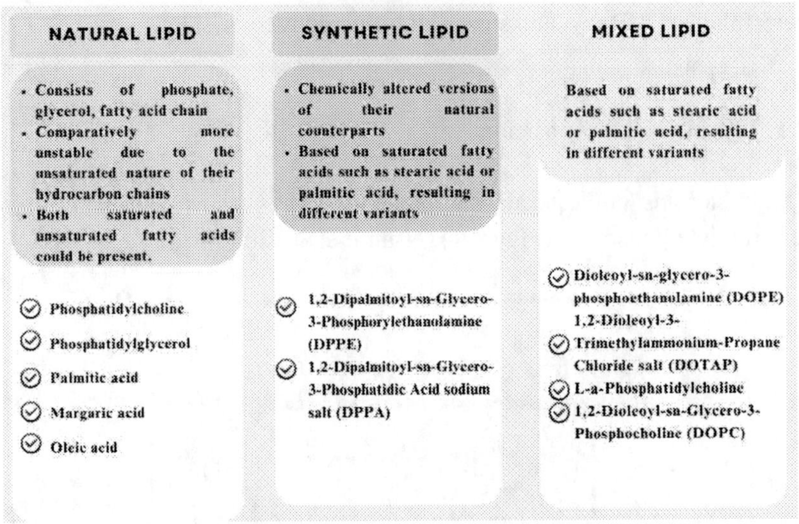

Figure 2. Classification of liquid components.

2.1.3. Other Lipid Additives

In addition to phospholipids and cholesterol, various lipid additives can be incorporated to tailor the properties of liposomes. For example, the inclusion of cationic lipids can enhance the encapsulation efficiency of negatively charged drugs (Allen and Cullis et al., 2013). Cationic liposomes can compact and enclose DNA or genetic materials within lipid nanoparticle (LNP) formulations, enabling their effective delivery into cells. The complexes formed by cationic liposomes and DNA exhibit an attraction toward the cell membrane with a negative charge, facilitating the transportation of hereditary data within a cell (Sun and Lu, 2023). Additionally, functional groups, such as polyethylene glycol (PEGylation), can be used for surface modification to improve drug delivery (Torchilin, 2005).

2.1.4. Surfactants

Surfactants have been employed in liposome formulations to alter the characteristics of encapsulation and discharge of the enclosed agent. Surfactants almost diminish the surface tension that exists between distinct immiscible phases. These amphiphiles, consisting of a single acyl chain, play the role of surfactants by disrupting the lipid bilayer present in liposomal nanoparticles. As a result, the deformability of these nanovesicles is enhanced (Nsairat et al., 2022). Surfactants frequently used in liposome composition consist of sodium cholate, Span 60, Span 80, Tween 60, and Tween 80 (Chen et al., 2017).

2.2. Surface Modification Strategies for enhanced drug delivery

2.2.1. PEGylation

PEGylation is a widely employed approach for the surface modification of liposomes, wherein polyethylene glycol chains are affixed to the surface. This modification creates a layer that protects the liposome surface, termed the "stealth effect," which reduces opsonisation, and uptake by the reticuloendothelial system (RES), and prolongs circulation time (Li and Huang, 2010). Lipid nanoparticles (LNP) with 1, 2-Dimyristoyl-Rac-Glycero-3-Methoxypolyethylene Glycol-2000 (DMG PEG-2000) are considered to be more stable (Shah et al., 2020).

2.2.2. Ligand Targeting

Liposomes could be modified by affixing ligands to their surfaces, enabling specific targeting to cells or tissues expressing corresponding receptors. Ligands such as antibodies, peptides, or aptamers can facilitate active targeting and improve drug delivery to desired sites (Marqués-Gallego and de Kroon, 2014).

2.2.3. Stealth Liposomes

Following intravenous administration, liposomes are frequently immediately eliminated from the bloodstream, resulting in a significant build-up of liposomes in the liver and spleen. This tendency poses a limitation to the therapeutic application of liposomes (Marqués-Gallego and de Kroon, 2014). Stealth liposomes or long-circulating liposomes are designed to evade detection by the immune system. Surface modification with PEG or other polymers provides a hydrophilic coating that prevents opsonisation and reduces interactions with plasma proteins, resulting in prolonged circulation and enhanced drug delivery (Immordino et al., 2006).

2.3. Factors Affecting Drug Load and Release from Liposomes

Factors influencing drug loading and release from liposomes include liposome composition, size, surface charge, and additional components (Figure 3; Table 3).

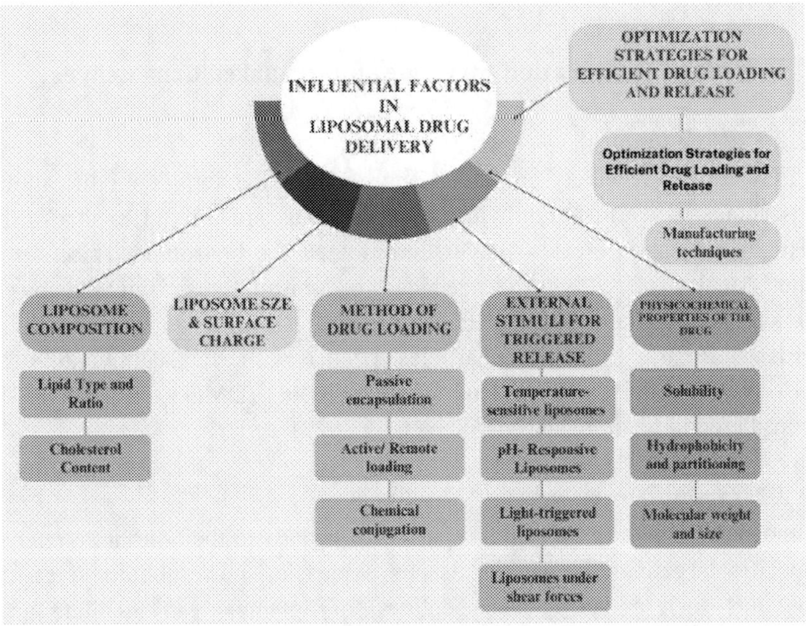

Figure 3. Influential factors in liposomal drug delivery system.

Researchers conducted a study in which they investigated the impact of various factors on black seed oil. They examined how the ingredients used and the methods of preparation influenced the performance in relieving pain in animal models. The researchers found that certain formulation attributes, such as the type and concentration of lipids, decide the stability and effectiveness of the liposomes. The investigation also delved into the implications of various process parameters, notable temperature, and mixing techniques on the quality of the liposomes. The results showed that increasing the proportion of oil in the liposomes reduced their size to a certain concentration. The efficiency of liposomes to capture the drug was influenced by the amount of cholesterol and category of cryoprotectant utilised during the generation process of the liposome. Stability tests revealed that the liposomal formulation remained stable under ordinary circumstances for 30 days (Rushmi et al., 2017).

Table 3. Factors that affect drug loading and release from liposomes

S. No.	Category	Factor	Affect and Outcome	References
1	*Liposome composition*	Influence of lipid types and ratios	Increasing the drug-to-lipid ratio enhances the release characteristics and alters the vesicle morphology, resulting in crystal formations and a significant increase in the half-life of doxorubicin release.	(Johnston et al., 2008)
		Impact of cholesterol content	Cholesterol, added at <30% of total lipids, provides rigidity and stability to liposomes by integrating into the structure of the lipid bilayer.	(Nsairat et al., 2022)
2	*Liposome Size and Surface Charge*	Effect of liposome size on drug loading and release	LipoplatinTM (110 nm) exhibits controlled release, influencing pharmacokinetics. Liposomes (50-200 nm) impact biodistribution, and rapid release mimics free drug kinetics but enhances solubility.	(Zhang et al., 2016; Allen and Cullis, 2004)
		Influence of Surface Charge on drug encapsulation and release	Negatively charged liposomes have reduced half-life, whereas positively charged liposomes can be toxic and rapidly cleared; lipoplex systems have applications in targeted DNA delivery.	(Immordino et al., 2006; Allen and Cullis, 2004)
3	*Method of Drug Loading*	Active-, Passive- Loading, Chemical Conjugation	Passive encapsulation involves hydrating a thin film with a drug-containing solution, while active loading utilises electrochemical or pH gradients for efficient drug uptake; chemical conjugation links drugs to liposomes.	(Vakili-Ghartavol et al., 2020; Sur et al., 2014; Allen and Cullis, 2004; Almeida et al., 2020)
4	*External Stimuli for Triggered Release*	Temperature-sensitive liposomes	A study on thermosensitive liposomes in tumour microvascular networks showed that mild hyperthermia enhances liposome extravasation, allowing targeted delivery of therapeutic agents.	(Gaber et al., 1996; Ta and Porter, 2013)
		pH-Responsive Liposomes	pH-sensitive liposomes can selectively release encapsulated molecules in response to acidic environments, enhancing cellular uptake and delivery. Different types of pH-sensitive lipids, such as DAPE, POPE, DOPE, and cardiolipin, have been investigated. The GALA peptide undergoes pH-dependent conformational changes, enabling efficient release of contents from liposomes. (DAPE: diacetylenic-phosphatidyl-ethanolamine, POPE: palmitoyl-oleoyl-phosphatidyl-ethanolamine, DOPE: dioleoyl-phosphatidyl-ethanolamine AND GALA: 30 amino acids, with a repeating sequence of glutamic acid-alanine-leucine-alanine)	(Alrbyawi et al., 2022; Chu and Szoka, 1994; Liu and Huang, 2013; Sasaki et al., 2008)

Table 3. (Continued)

S. No.	Category	Factor	Affect and Outcome	References
		Light-triggered liposomes	Light-triggered liposomes utilise different mechanisms, such as lipid oxidation, photocross-linking, photoisomerization, photocleavage, and photothermal release, for controlled drug release. An example of a successful light-activated treatment is Visudyne.	(Miranda and Lovell, 2016; Pidgeon and Hunt, 1983; Subramaniam et al., 2010; Arias-Alpizar et al., 2020)
		Liposomes under shear forces	Under shear, liposomes experience changes in their structure and dynamics, which can affect their stability and drug-release properties (Figure 4).	(Karaz and Senses, 2023)
5	*Physicochemical Properties of the Drug*	Solubility considerations, Hydrophobicity & partitioning, and Molecular weight & size	Solid pro-liposomes and freeze-dried liposomes improve drug solubility and prevent crystallisation during storage. Liposomes containing distearoyl phosphatidylcholine as the main phospholipid was found to be more stable than those containing dipalmitoyl phosphatidylcholine. The partition of drugs into liposomes can provide information on their potential absorption and bioavailability in the gastrointestinal tract. Smaller liposomes tend to exhibit longer circulation times and enhanced tissue permeation in contrast to larger liposomes.	(Lee, 2020; Khan et al., 2008; Balon et al., 1999a; Balon et al., 1999b; Sercombe et al., 2015)
6	*Optimisation Strategies for Efficient Drug Loading and Release*	Formulation approaches	Second-generation liposomes with modified lipid composition and surface, incorporating ligands, enable targeted and sustained drug delivery, improving disease control and minimising adverse effects in cancer therapy. Research deciphers the development of d-a-Tocopheryl Polyethylene Glycol 1000 Succinate liposomes conjugated with trastuzumab for the targeted and sustained delivery of docetaxel to cancer cells expressing human epidermal growth factor receptor 2 and reduces the after-effects.	(Khan et al., 2020; Raju et al., 2013)

S. No.	Category	Factor	Affect and Outcome	References
		Use of co-solvents and co-surfactants	(Figure 5)	
7	*Manufacturing techniques*	Methods of drug loading in liposomes and types of liposomes	The conventional methods for manufacturing liposomes involve four main stages: lipid drying, lipid dispersion in water, purification of liposomes, and analysis of the final product. However, these methods have limitations in encapsulating water-soluble agents. The variety of liposomes includes oligolamellar morphology refers to vesicles with two to five concentric lamellae, whereas multilamellar vesicles are liposomes with more than five lamellae. Fusogenic liposomes and antibody-mediated liposomes are employed in cancer therapy. (Figure 6 and Figure 7)	(Mozafari, 2005; Moghimi, 2012; Sharma et al., 2018)
		Liposome Preparation Methods	*Lipid Film Hydration is* a physical dispersion method also known as the solvent evaporation technique and called the Bangham method of liposome formation. (Figure 8 and Figure 9)	(Shashi et al., 2012; Shah et al., 2020; Lombardo and Kiselev, 2022)
			The hand-shaken method encompasses the dissolution of the lipid mixture and charged components in an organic solvent, evaporating the solvent, and hydrating the lipid residue with a phosphate buffer.	(Shashi et al., 2012)
			The nonshaking method closely resembles the hand-shaken method but involves carefully swelling the lipid film by adding a bulk liquid (e.g., sucrose solution).	
			Freeze drying or lyophilization is an alternative method in which lipids that are dissolved in an organic solvent undergo freeze drying, and the resulting lipid powder is then hydrated with an aqueous medium. Slow freezing rates minimise super-cooling and osmotic pressure, reducing ice crystal formation within the liposome's interior compartment and preventing drug leakage (Figure 10).	(Shashi et al., 2012; Franzé et al., 2018)

Table 3. (Continued)

S. No.	Category	Factor	Affect and Outcome	References
8		Solvent Dispersion Methods	The ethanol injection method allows the formation of multilamellar vesicles (MLVs) with micrometer-sized particles. This approach is best suited for lipophilic medicines, which have high entrapment efficiency. (Figure 11) Ether injection involves slowly injecting an immiscible organic solution into an aqueous phase at the vaporisation temperature of the organic solvent. The reverse-phase evaporation method shows remarkable entrapment efficiency. This method can passively entrap hydrophilic drugs up to 30-50%, whereas the active loading technique can reach entrapment of more than 90%. Although suitable for small-volume parenterals, its scale-up is restricted due to the intricate nature of the manufacturing procedure (Figure 12).	(Shah et al., 2020; Shashi et al., 2012)
9		Detergent Solubilization Technique	This method involves bringing phospholipids into contact with an aqueous phase through detergents, forming micelles that associate with phospholipids. As the concentration of detergent increases, the micelles reduce in size.	(Shashi et al., 2012)
10	Processing Techniques for Hydrated Lipids through the above methods	Micro-emulsification, sonication, membrane extrusion	Microemulsification uses a microfluidizer to create small vesicles from a lipid suspension, while sonication reduces the size of the vesicle and imparts energy. Membrane extrusion passes liposomes through a filter (for instance 0.2 μm or greater) to reduce size, and freeze and thaw sonication involves rupturing and reforming vesicles through freeze-thaw cycles.	(Shashi et al., 2012; Shailesh et al., 2009; Ong et al., 2016)
11	Purification of Liposomes	Gel filtration chromatography, dialysis, centrifugation	Liposomes can be purified using gel filtration chromatography, dialysis, or centrifugation, with Sephadex-50 commonly used for chromatography, hollow fibre dialysis cartridges for dialysis, and different centrifugation speeds for separating SUVs and MLVs. Detergent removal methods include dialysis, gel permeation chromatography, and dilution, using beads such as Sephadex G-50, G-100, Sepharose 2B-6B, and Sephacryl S200-S1000.	(Shashi et al., 2012; Sharma et al., 2018)

S. No.	Category	Factor	Affect and Outcome	References
12	Size manipulation		The manipulation of liposome size is an essential aspect of drug delivery. Numerous methodologies including sonication, freeze-thaw, homogenisation, and extrusion are employed in this regard, each with its own set of merits and demerits. It is widely acknowledged that extrusion is the most reliable and effective method.	(Shah et al., 2020)
13	Evaluation of liposomes:		The liposome is studied for properties like appearance, size, size distribution, surface charge, percent drug encapsulation, vesicle shape, lamilarity, assay, *in vitro* release, and stability.	(Sharma et al., 2018; Shah et al., 2020)

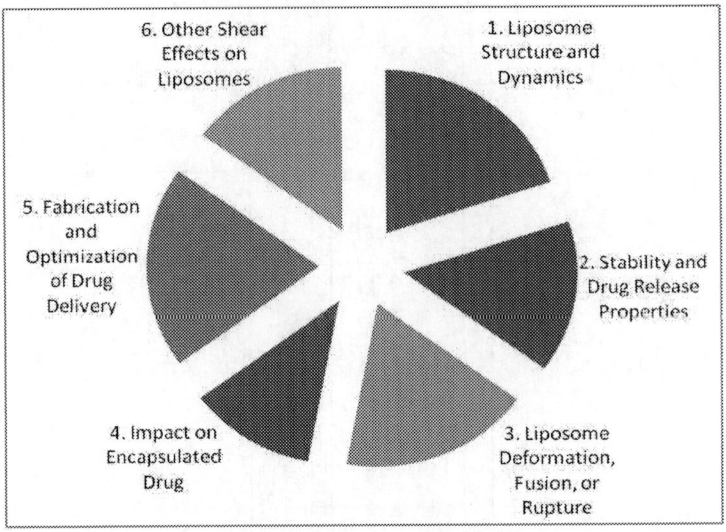

Figure 4. Illustration of the distribution of the effects of shear on liposomes.

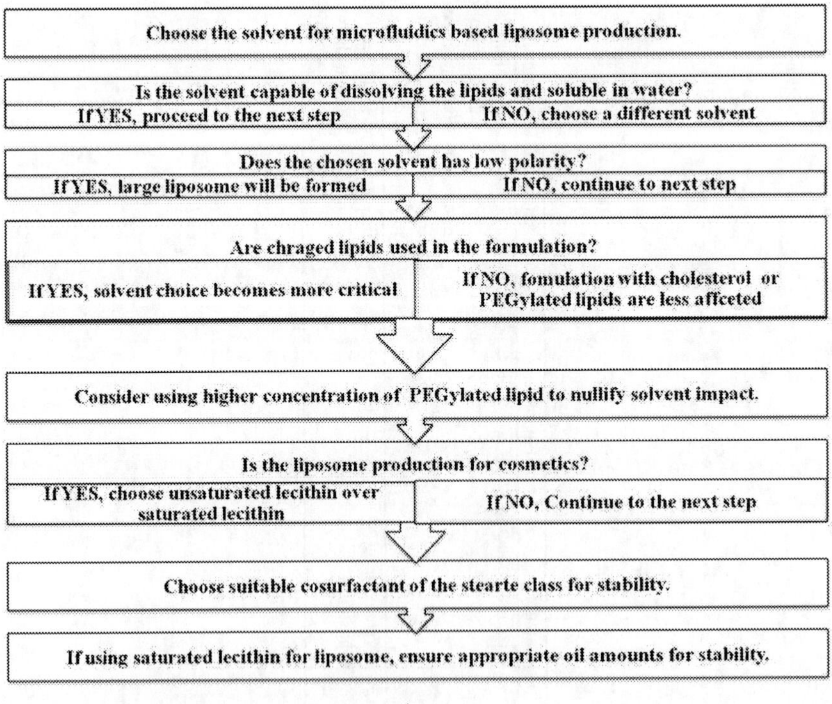

Figure 5. Representation of the decision-making process for solvent selection in microfluidics-based liposome production.

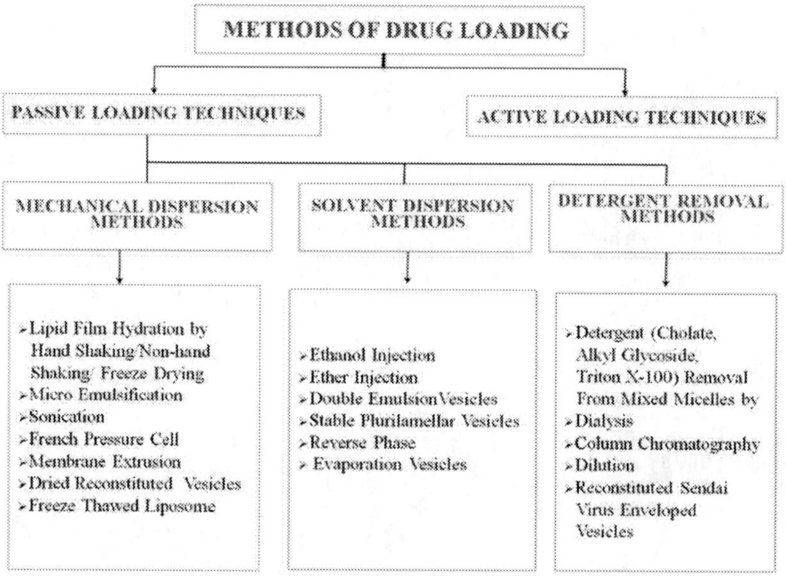

Figure 6. Methods of drug loading in Liposome.

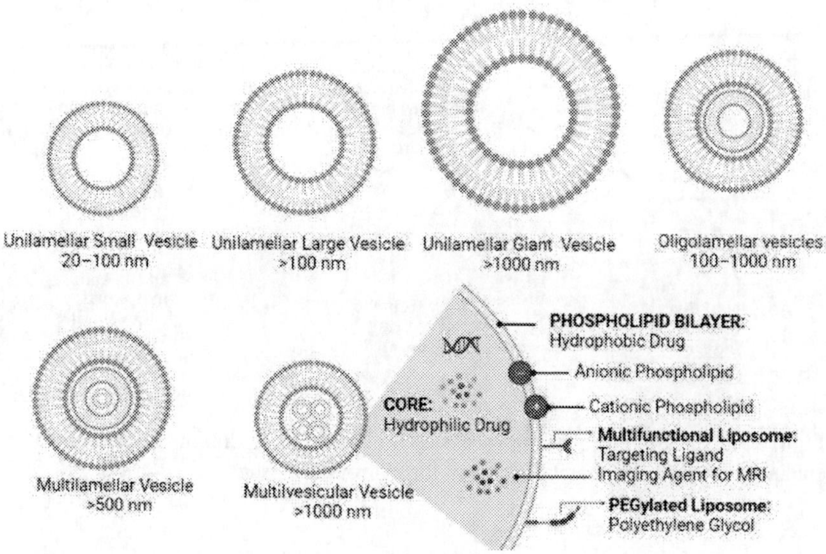

Figure 7. Types of liposome and Surface Modifications.

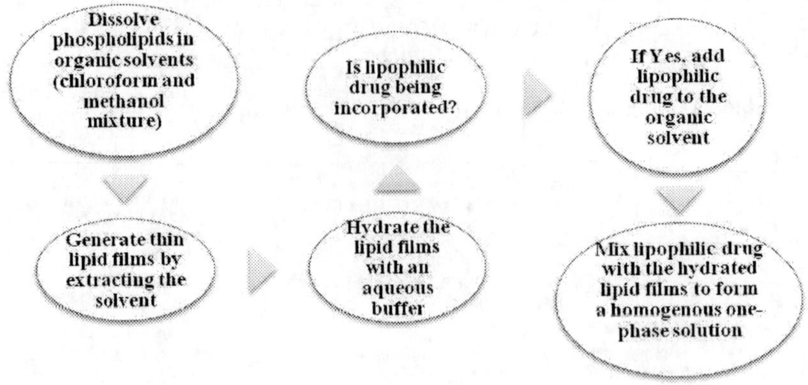

Figure 8. Liposome formation process using the lipid film hydration method for lipophilic drugs.

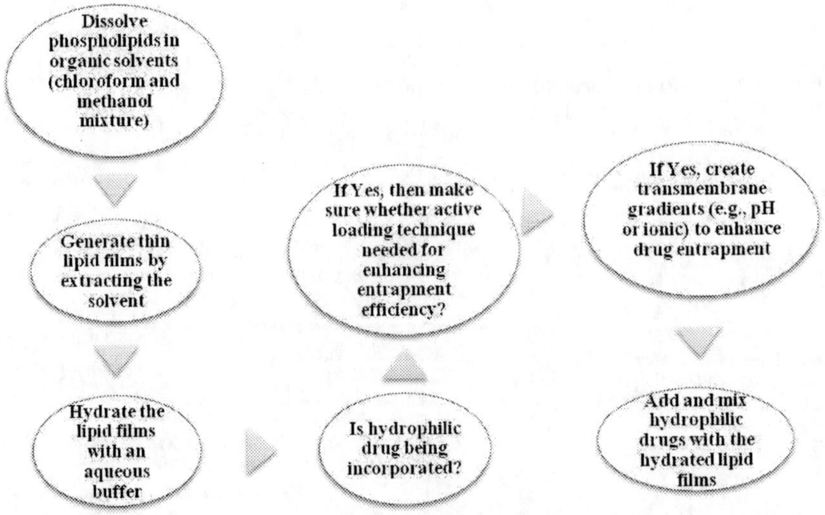

Figure 9. Liposome formation process by the lipid film hydration method and enhancement of entrapment efficiency for hydrophilic drugs.

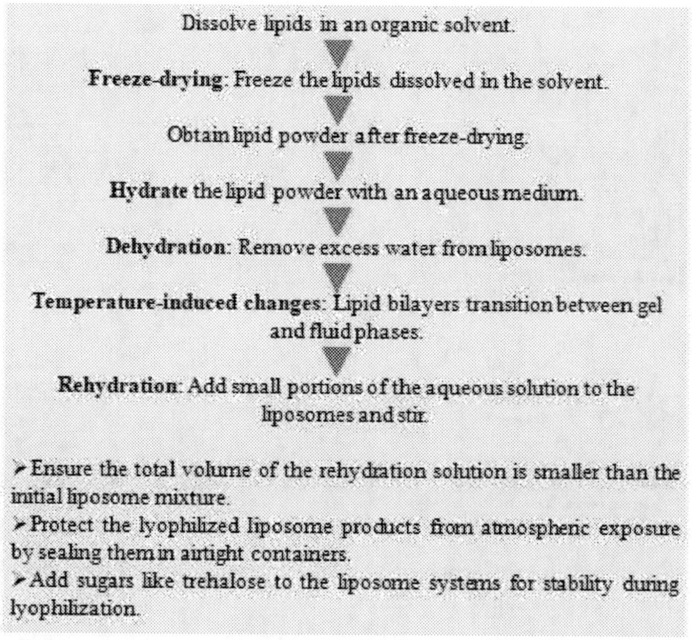
Dissolve lipids in an organic solvent.
Freeze-drying: Freeze the lipids dissolved in the solvent.
Obtain lipid powder after freeze-drying.
Hydrate the lipid powder with an aqueous medium.
Dehydration: Remove excess water from liposomes.
Temperature-induced changes: Lipid bilayers transition between gel and fluid phases.
Rehydration: Add small portions of the aqueous solution to the liposomes and stir.
➢Ensure the total volume of the rehydration solution is smaller than the initial liposome mixture.
➢Protect the lyophilized liposome products from atmospheric exposure by sealing them in airtight containers.
➢Add sugars like trehalose to the liposome systems for stability during lyophilization.

Figure 10. Step-by-step lyophilization process of liposomes.

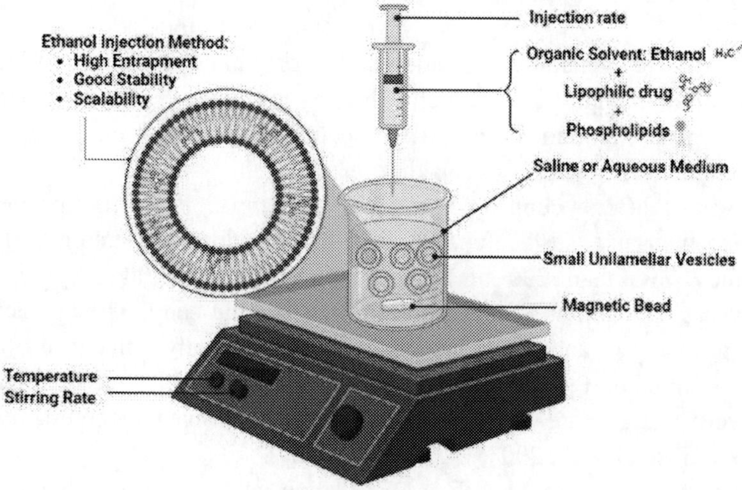

Figure 11. Illustration of the ethanol injection method used for the preparation of liposomes.

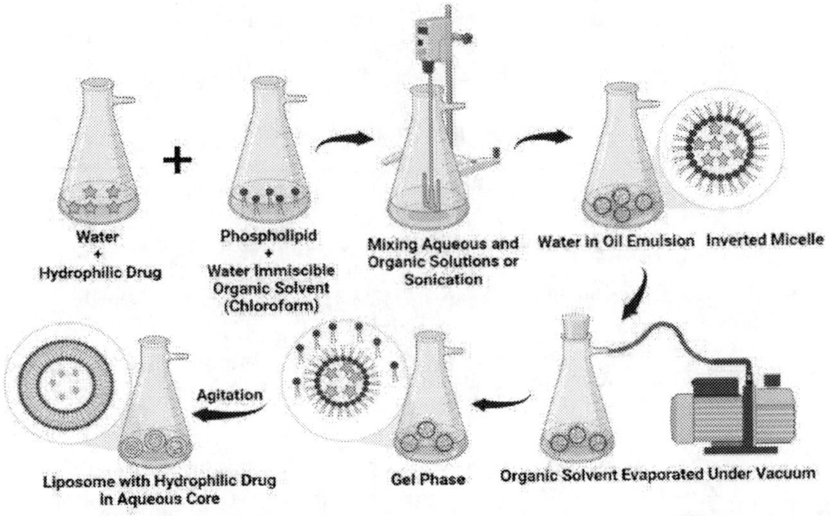

Figure 12. Schematic representation of the reverse phase evaporation method utilised for the preparation of liposomes.

2.4. Biocompatibility and Toxicity Considerations of Liposome Formulations

Liposomes are generally considered biocompatible because of their phospholipid composition, which closely resembles cell membranes. However, it is important to evaluate its potential toxicity to ensure its safe use in biomedical applications (Nakhaei et al., 2021).

Several factors contribute to the biocompatibility of liposomes, including their size, surface charge, lipid composition, and stability. Small liposomes (less than 200 nm) are generally preferred as they exhibit improved circulation and reduced clearance from the body. The interaction between liposomes and biological systems is significantly influenced by the surface charge of the liposomes. Liposomes that are neutral or slightly negatively charged are more likely to exhibit improved biocompatibility profiles (Krasnici et al., 2003; Inglut et al., 2020).

Surface modifications, such as PEGylation (attachment of polyethylene glycol), can improve the biocompatibility of liposomes by minimising their recognition and clearance by the immune system (Suk et al., 2016; Inglut et al., 2020).

Toxicity assessment involves studying the effects of liposomes on cells, tissues, and organs. Various *in vitro* and *in vivo* models are used to evaluate the cytotoxicity, immunogenicity, and potential adverse effects of liposome formulations. It is important to consider factors such as leakage of encapsulated drugs, accumulation in organs, and potential inflammatory responses (Inglut et al., 2020).

A study by (Knudsen et al., 2015) evaluated the *in vivo* toxicity of cationic micelles and liposomes, revealing significant toxicity and adverse effects. Cationic nanoparticles exhibited higher toxicity compared to neutral or anionic nanoparticles. Caution is necessary when using cationic micelles and liposomes in drug delivery.

Researchers discussed the evaluation of the toxicity of liposome-encapsulated sirolimus through *in vivo* and *in vitro* research. *In vivo*, the liposomal formulation showed minimal toxicity and no significant adverse effects. *In vitro*, liposomal sirolimus had lower cytotoxicity compared to free sirolimus, indicating improved biocompatibility (Abud et al., 2019).

Conventional techniques used for liposome formation possess a defect in which solvent residue exceeds the FDA threshold, resulting in toxicity. However, the advanced technology Superlip process offers a solution by employing a low solvent residue and a green process that employs carbon dioxide, a non-toxic substance (Trucillo et al., 2020).

As a result of their excellent biocompatibility and low toxicity, liposomes have emerged as a significant group of therapeutic drug nanocarriers approved for clinical use in cancer treatment (Lombardo and Kiselev, 2022).

3. Recent developments in Liposome Technology

Investigators developed a drug delivery system that aimed to overcome the limitations of anticancer drugs, such as low permeability and degradation (Sharma et al., 2018). Recent advancement in liposome technology has resulted in the emergence of more sophisticated and efficient liposome-mediated pharmaceutical delivery systems. These developments include the use of novel lipids, surface modifications, and targeting moieties. Novel lipids, such as PEGylated lipids, have been used to enhance the stability and biocompatibility of liposome formulations. Surface modifications, such as the attachment of antibodies or peptides, have been used to target liposomes to specific cells or tissues. Targeting moieties, such as folate or transferrin, have also been used to target liposomes to cancer cells (Figure13) (Torchilin, 2005).

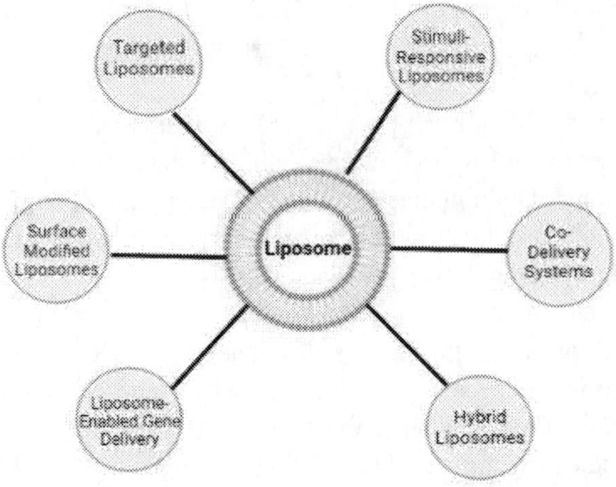

Figure 13. Categories of some recent developments in liposome-mediated delivery.

Fusogenic liposomes composed of the Sendai virus combined with conventional liposomes were developed to deliver drug contents directly into the cytoplasm using fusion mechanisms. These liposomes bypassed endocytosis and showed potential as effective drug carriers. Additionally, antibody-mediated or ligand-mediated interactions have been explored as next-generation drug carriers for targeted delivery to cancer cells (Sharma et al., 2018).

3.1. Heating Method

A solvent-free heating method for fast liposome production has been discovered. An isoniazid-conjugated phthalocyanine liposome using hydration has been researched that yielded high entrapment efficiencies. Another research prepared surfactant vesicles that encapsulate alpha-tocopherol with a high encapsulation efficiency using a modified heating method. This is a scalable method based on green chemistry principles that is suitable for the encapsulation of hydrophilic and lipophilic drugs (Figure 14) (Shah et al., 2020; Mozafari, 2005).

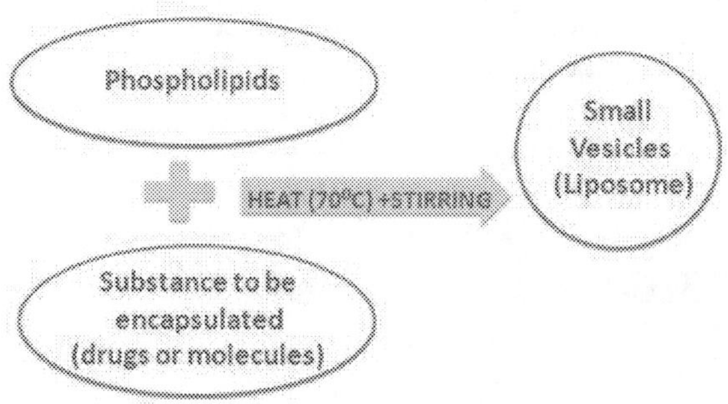

Figure 14. Heating Method for Liposome Formation.

3.2. Combining Nanoprecipitation with Ionic Interaction

A vesicular system called LeciPlex® was developed using a single-step fabrication method. To improve the characteristics of the phospholipids, researchers mixed them with a charge-imparting agent and dissolved them in a nontoxic solvent named transcutol HP. After the solvent phase to thermal energy above the lipid phase transition temperature, it was subsequently hydrated utilising an aqueous medium that was concurrently maintained at the same temperature. The result was the development of self-organised vesicles with nanometer-scale dimensions and exceptional stability during storage (Figure 15). The LeciPlex® technique has been used effectively to encapsulate various drugs and genetic materials, demonstrating its versatility and wide applicability (Shah et al., 2020).

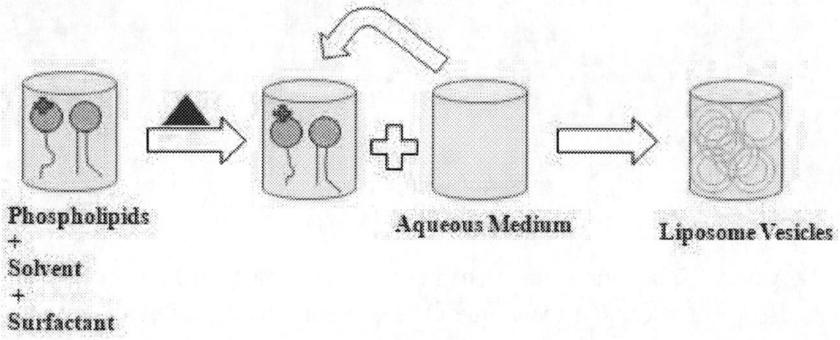

Figure 15. Schematic representation of Nanoprecipitation combined with ionic interaction corresponding to LeciPlex®.

3.3. Novel Apparatus for Liposome Fabrication Based on Rapid Solvent Exchange

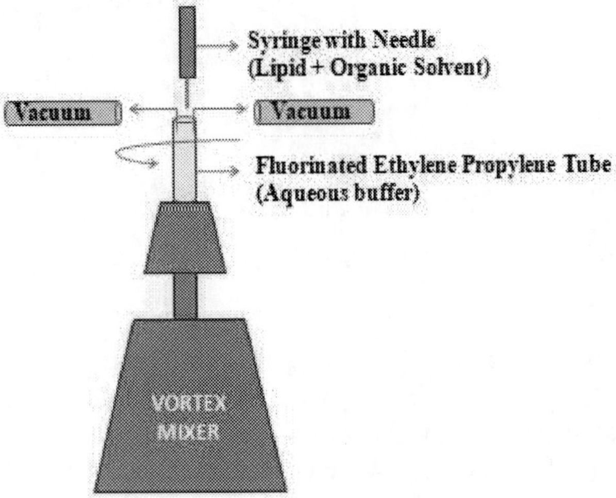

Figure 16. Schematic representation of Solvent Exchange Apparatus.

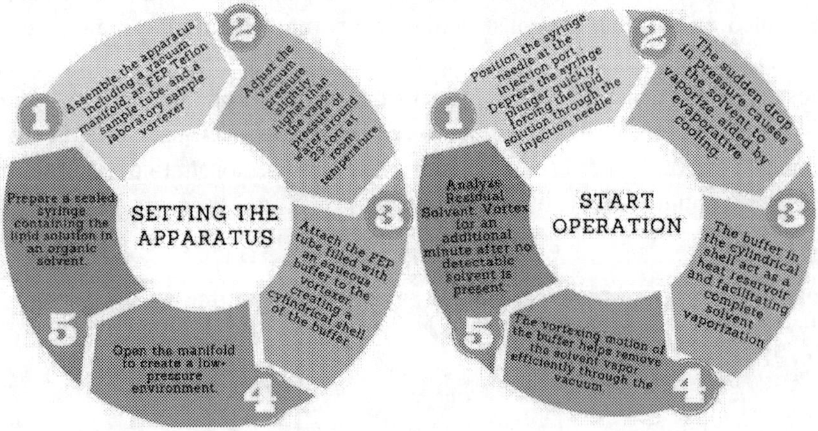

Figure 17. Sample preparation process using the rapid solvent exchange method.

A novel apparatus is developed for the fabrication of liposomes using a rapid solvent exchange technique. The apparatus is a tube with a vortexer and a buffer, which allows controlled vaporization of the solvent (Figure 16). It is reported that liposome production is significantly influenced by mechanical forces in vortex mixing and flow rates. High vortex speed generated rapid evaporation resulting in ultra-large vesicles (ULVs) formation while reducing the vortex speed formed multilamellar vesicles (MLVs). Factors such as the amount of lipids, the organic solvent-to-water ratio, and the presence of buffers influenced the assembly of ULVs or MLVs. The preparation time for the liposomes using this apparatus was approximately four minutes (Figure 17) (Shah et al., 2020).

3.4. High-Shear Method for Liposome Preparation

This method was patented for immunogenic liposomes containing vaccine adjuvants. The impact of intense shear on a surfactant solution that holds MLVs results in the production of ULVs. MLVs are transformed into ULVs within the size range of 50-75 nm by homogenising at 200 bar, but an enlargement in size occurs during storage (Figure 18) (Shah et al., 2020).

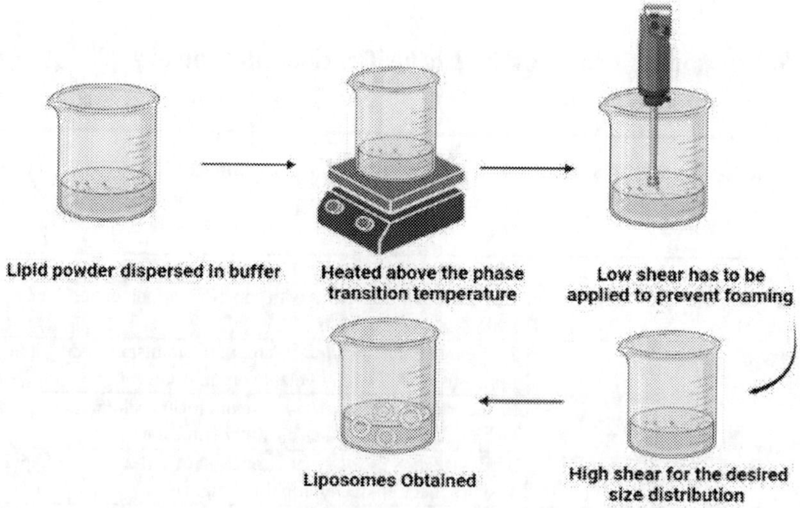

Figure 18. High Shear Method for liposome preparation.

3.4.1. Shearing with Glass Beads

Glass beads are used to generate shear forces and reduce the size of liposomes. The size of the liposomes varies with the diameter of the glass beads used (Figure 19) (Shah et al., 2020).

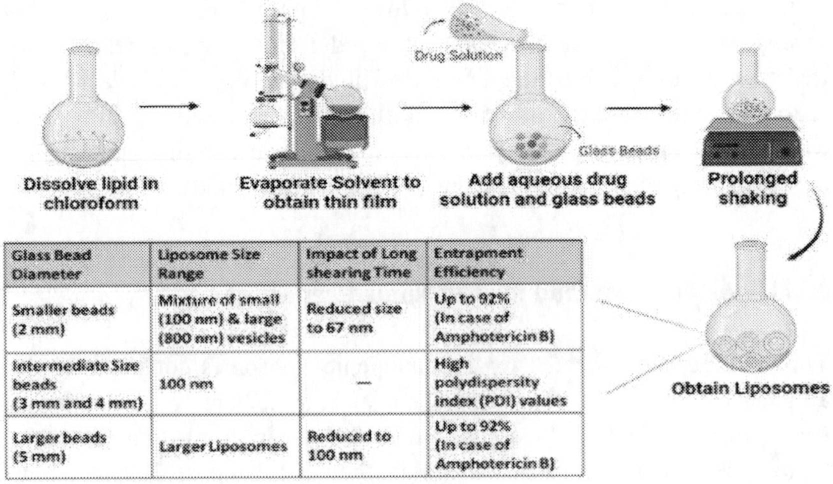

Figure 19. Liposome preparation with glass beads.

3.5. Liposome Formation by Emulsification and Solvent Evaporation Technique

Table 4. Aspects of liposome formation by emulsification with solvent evaporation technique

Aspects	Multiple emulsification-solvent evaporation
Goal	Preparation of liposomes with superior entrapment efficiency for hydrophilic components.
Process	*1. Primary emulsification:* Water in oil emulsion of drug and lipids suspended within volatile organic solvent.
	2. Secondary emulsification: The microchannel emulsification technique results in water/oil/water emulsions.
	3. Evaporation of solvents: Formation of self-assembled lipid vesicles that trap the hydrophilic drug.
Control over size	Achieved through the size of primary emulsion droplets and the emulsification technique used.
Size range of vesicles	0.2 μm to several microns.
Factors affecting entrapment efficiency	Droplet size and surfactant type.
Multilamellar vesicles (MLVs)	Can also be produced by this technique.
High entrapment efficiency	Microchannel emulsification with sodium caseinate as an emulsifier showed entrapment efficiency of up to 82%.
Advantages of micro-channel emulsification	1. Absence of heating and low shear.
	2. The mitigation of the transfer of hydrophilic substances from the internal phase to the external phase.

Liposome formation by the emulsification and solvent evaporation technique involves the creation of water/oil/water emulsions subsequently to the evaporation of the solvent, resulting in self-assembled lipid vesicles with high entrapment efficiency for hydrophilic moieties (Table 4) (Shah et al., 2020).

3.6. Packed Bed-Based Reactors

A packed bed reactor that rotates for continuous liposome production utilises high centrifugal force, controlled flow rates, and temperature regulation to influence the size of the liposome and drug entrapment efficiency. Factors such as the flow rate ratio (FRR) and the high gravity level (HGL) play crucial roles in determining particle size and drug loading. It shows potential for large-scale liposome manufacturing with optimised conditions achieving high production output and desirable liposome characteristics (Shah et al., 2020).

3.7. PVA Coating for Solid-State Liposome Formation

A recent approach to developing a liposome involves the application of polyvinyl alcohol (PVA) coating on coverlips, followed by lipid spreading, solvent evaporation, and the observation of liposome formation by microscopy. The technique was also used for protein encapsulation. The formation of GUV was observed through microscopy (Shah et al., 2020).

3.8. Spray drying for Liposome Preparation

A one-step spray drying process has been developed to prepare liposomes. This process entails the dissolution of lipids, drug, and lactose in a solvent system and spray drying it (Maniyar and Kokare, 2019). Gala et al., in 2015 developed a method using fluid bed coating, high-pressure homogenization, and lyophilization. Proliposomes were prepared by spraying an ethanolic solution onto sucrose particles, followed by hydration and freeze-drying. The resulting liposomes were nanosized and freeze-drying retained their size below 155 nm (Table 5) (Gala et al., in 2015).

Table 5. Summary of a recent report on the spray drying method for liposome preparation

Liposome Preparation Method	Size (nm)	Polydispersity Index (PDI)	Entrapment Efficiency (%)	Reference
Spray Drying	270	0.239	56.38%	(Maniyar and Kokare, 2019)
Fluid Bed Coating + High-Pressure Homogenization + Lyophilization	70-125	Not specified	Increased if drying time 2 hours to evaporate excess ethanol	(Gala et al., 2015)

3.9. Solvent Diffusion Methods

Researchers have developed modified methods to simplify liposome production (Table 6).

Table 6. Modified solvent diffusion methods for liposome preparation

Researcher & Year	Method	Process Parameter & Liposome Characteristics
Costa et al., 2016	Modified Ethanol Injection	Utilises pressurised tanks and a static mixer for improved mixing. The size may differ based on lipid type and flow rate.
Araki et al., 2018	Inline Method with Thermal Mixing and Dialysis	It involves lipid dissolution, mixing, and self-assembly to form liposomes. The process resulted in monodisperse liposomes without additional homogenisation. The liposome size is influenced by adjusting the amount of organic solvent. Scalable, aseptic, and automatable process.

3.10. Supercritical Fluids

Supercritical fluids (SCFs) are gaining popularity as a substitute for organic solvents because of their ability to enhance separation and purification processes more effectively (Patil and Jadhav, 2014).

Supercritical fluids have been utilised in liposome preparation through a process called SuperLip (Lesoin et al., 2011). In a study, an ethanolic lipid solution was mixed with CO_2 to make an expanded fluid, which was then combined with an atomised aqueous solution of the drug leading to the formation of liposome suspension. A subsequent decompression step allows recovery of the liposomes (Table 7) (Trucillo et al., 2017; Trucillo et al., 2019).

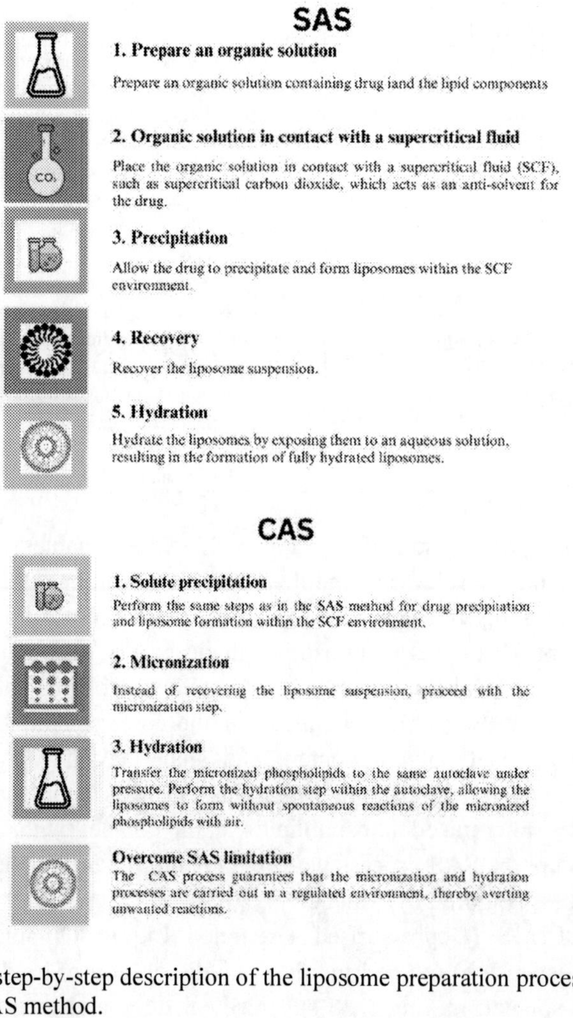

Figure 20. A step-by-step description of the liposome preparation process using the SAS & CAS method.

The RESS technique (Rapid Expansion of Supercritical Solution) involves dissolving lipids in ethanol and supercritical CO_2 followed by rapid expansion through a nozzle (Table 7) (Lombardo and Kiselev, 2022).

Table 7. Supercritical Fluids for Liposome Preparation: Techniques and outcomes

Technique	Liposome Preparation Method	Outcomes/Applications
SuperLip	Lipids dissolved in ethanol mixed with CO_2	• Elevated level of entrapment efficiency concerning both hydrophilic and lipophilic drugs. • Encapsulation efficiency is dependent on the flow rate of the aqueous solution. • Smooth, homogeneous liposomes with sizes ranging from 300 to 600 nm. • Successful preparation of liposomes with antioxidants (Trucillo et al., 2017; Trucillo et al., 2019).
RESS (Rapid Expansion of Supercritical Solution)	Lipids dissolved in ethanol and supercritical CO_2	• Formation of liposomes capable of encapsulating hydrophobic and hydrophilic drugs. • Control over liposome size and encapsulation efficiency through processing conditions. • Successful encapsulation of essential oils, melatonin, and vitamin C in liposomal nanocarriers (Lombardo and Kiselev, 2022).

The supercritical antisolvent approach (SAS) enables lower residual solvent content, is a relatively simple process, and can be used for molecules with poor solubility in the supercritical fluid (SCF) because the SCF acts as an antisolvent. Examples of liposomes made by this technique are vitamin D3 (VD3) pro-liposomes and DSPC-PEG unilamellar liposomes loaded with albendazole. The liposomes obtained via the SAS technique shall have a diminished residual solvent content and shall lack any sign of organic solvents. To overcome certain limitations, such as spontaneous reactions exhibited by micronised phospholipids upon contact with air before the hydration step, the SAS process was upgraded to the continuous anti-solvent (CAS) process (Figure 20) (Maja et al., 2020).

The DELOS (Depressurised Expanded Liquid Organic Solution in Aqueous Suspension) method involves the dissolution of lipids and drugs in an organic solvent, mixing with supercritical CO_2, and then expanding the mixture to CO_2 (Figure 21). The resulting bubbles transport lipids to water, leading to the formation of liposomes. DELOS offers advantages such as small and uniform liposome size, stability, and simplified manufacturing (Lombardo and Kiselev, 2022). The main benefit of the DELOS method is that it allows one to work with thermosensitive materials and manufacture fine particles without melting them. Furthermore, the DELOS process only requires a small amount of working pressure, such as 10 MPa at 35°C, which is an important factor from a financial point of view, particularly for industrial use (Maja et al., 2020). Recent applications include the development of pH-sensitive lipid nanovesicles for miRNA delivery and α-galactosidase-loaded nanoliposomes loaded with -galactosidase for the treatment of Fabry disease (Lombardo D et al., 2022).

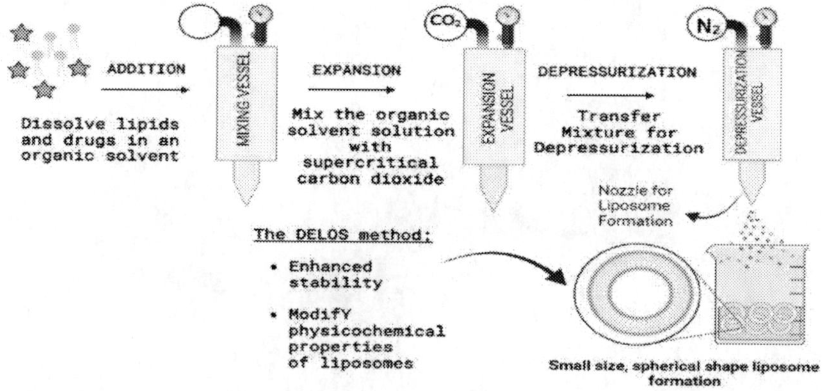

Figure 21. Schematic representation of liposome formation using the DELOS method.

3.11. Microfluidics Approach

The microfluidic approach is gaining attention as an advanced method for producing nanoparticles, especially in the field of nanomedicine, to customize the size distribution of liposomes. The traditional method requires additional steps to adjust liposome size, but this method eliminates the need for post-processing (Table 8) (Streck et al., 2019).

Table 8. Summary of microfluidic approach for liposome formation

Aspect	Microfluidics Approach
Mixing Technique	Rapid and controlled mixing of liquid reagents in small volumes.
Channel Size	Typically ranges from 5 to 500 micrometres.
Nanoparticle Formation	This was achieved by changing the flow rate ratio between alcohol and aqueous streams.
Influencial Factors	The width and mixing of the alcohol stream have a greater influence on liposome formation than the forces at the solvent-buffer interface.
Application	Suitable for producing both polymer and lipid-based nanoparticles used in nanopharmaceuticals.

3.11. Pulsed Jet-Flow Microfluidic Method

The pulsed jet flow microfluidic method involves one limitation which is the difficulty of automating the process, as the positioning of the microcapillary near the bilayer requires manual control (Figure 22) (Lombardo and Kiselev, 2022).

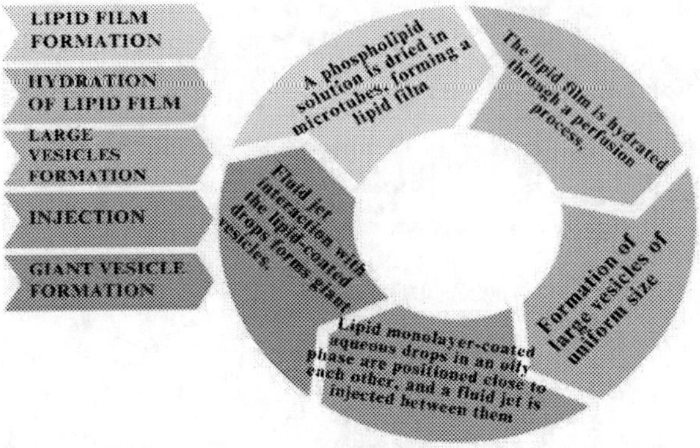

Figure 22. This diagram represents the sequence of steps involved in the pulsed jet flow microfluidic method, starting from the lipid film formation, followed by hydration to form liposomes, and ultimately resulting in the formation of giant vesicles.

3.12. Membrane Contactor

The membrane contactor technique represents a modified approach to the ethanol injection method for making liposomes (Table 9; Figure 23) (Lombardo and Kiselev, 2022).

Details of some new liposome formulations are discussed in Table 10.

Table 9. Various aspects of membrane contactor technique

Aspect	Membrane Contactor Technique
Method	Pressurised vessels with organic and aqueous phases separated by a porous glass membrane.
Liposome Formation	The lipid phase undergoes pressurisation through the membrane, transitioning into the aqueous phase, leading to self-assembled liposomes.
Advantages	High encapsulation efficiency, and precise control over liposome size.
Potential	Scalability for extensive manufacturing, and continuous production of nanosized liposomes.
Experimental Results	Successful liposome formation with different drugs and stable liposome suspensions.

Table 10. Current status of research and development in the formation of novel liposomes

Novel Liposome formulations	Aspects	Information	Reference(s)
Onpattro®	What is Onpattro® liposomal therapy?	It is a recently approved liposomal therapy for transthyretin-mediated amyloidosis (ATTR), using antitransthyretin small interfering RNA (siRNA) of antitransthyretin as a therapeutic agent.	(Urits et al., 2020)
	What is the composition of this formulation?	Onpattro® is made up of phospholipid, cholesterol, ionisable cationic lipids, and PEG2000-C-DMG (PEGylated myristoyl diglyceride).	
	What is the process of its formation?	They are formed through rapid mixing under acidic pH and undergo fusion into larger lipid nanoparticles as the pH neutralises.	
	What is the mechanism of action of Onpattro®?	In the body, liposomes target the endosome. At low pH, the cationic lipids become activated and prompt endosomal rupture, ultimately resulting in the discharge of siRNA into the cytosol.	
	What does the clinical trial suggest about Onpattro®?	Clinical trials have shown reduced transthyretin production and improved symptoms compared to placebo, with manageable side effects. Other ongoing trials focus on encapsulating approved drugs within liposomes, particularly for chemotherapeutics and siRNA-based therapies.	
	What are the benefits of Onpattro®?	Precise delivery, greater efficacy, and less systemic toxicity.	
LipocurcTM	What is liposomal therapy LipocurcTM?	LipocurcTM, a liposomal formulation of chemotherapeutics, is being investigated for advanced cancer.	(Almeida et al., 2020)
AtuPLEXTM	What is the AtuPLEXTM liposomal therapy?	AtuPLEXTM, liposomes loaded with small interfering RNA, are currently being investigated as a novel therapeutic strategy for conditions such as progressive pancreatic carcinoma and solid malignancies.	
Liposome based vaccines	Is liposome-based vaccines available?	Lipid nanoparticles utilised by Moderna and Pfizer for COVID-19 vaccines; Liposome-based vaccines showing promise in combating SARS-CoV and SARS-CoV-2	(Agarwal, 2022)

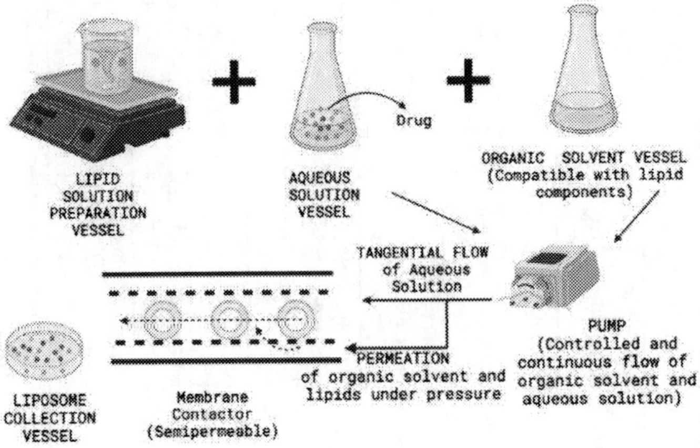

Figure 23. Schematic diagram of the formation of liposomes via membrane contactor method. (Lombardo and Kiselev, 2022).

4. Challenges Associated with Liposome-Mediated Drug Delivery

Challenges associated with liposome-mediated drug delivery still exist notably low efficiency of drug encapsulation, limited stability, and suboptimal bio-distribution. Other obstacles associated with liposome development and commercialisation include such as individual variations in enhanced permeability and retention, faster blood clearance, pilot scale batch, batch uniformity, and excipient control (Table 11) (Liu et al., 2022).

4.1. Challenges

4.1.1. Low Drug Encapsulation Efficiency

Encapsulation efficiency is an important quality parameter for liposome products as it determines the proportion of medication successfully encapsulated within the liposomes to the entire amount of drug. The high entrapment efficiency ensures that most of the drug is contained within the liposomes, reducing the free drug present in the formulation. Free drug in a liposomal formulation can potentially lead to toxicity or undesirable distribution of the drug in the body. Therefore, maximising the efficiency of encapsulation is critical to boosting the effectiveness and safety of liposome-mediated drug delivery (Bian et al., 2022).

Encapsulation efficiency is directly influenced by the size of the liposome and its distribution. Smaller liposomes have the advantage of crossing biological membranes. However, this can have implications for encapsulation efficiency. Smaller liposomes may have a reduced capacity to accommodate a large amount of drug due to their limited internal volume. As a result, the amount of drug that can be loaded into smaller liposomes may be lower compared to the size of larger liposomes. Additionally, reducing the size of liposomes also affects their stability. Liposomes that exhibit smaller dimensions possess a significantly greater ratio of surface area to volume. This leads to increased surface energy and thereby decreased liposome stability, potentially leading to drug leakage or aggregation (Nakhaei et al., 2021; Xia et al., 2014).

4.1.2. Limited Drug Stability

- Physical Instability

To achieve a more suitable thermodynamic state during storage, it has been discovered that liposomes have a propensity to combine and expand significantly in size. This phenomenon can lead to the drug escape from the vesicles, thereby jeopardising the physical stability of the liposomal drug product. Hence, it is imperative to measure the critical parameters such as vesicle shape, and size to gauge their physical stability. To carefully investigate the morphology and determine the size of the vesicles, light scattering, and electron microscopy can be opted for (Sharma et al., 2018; Nakhaei et al., 2021).

- Chemical Instability

Liposomes are prone to chemical breakdown, leading to changes in their composition and structure over time. This degradation can occur through various mechanisms, such as hydrolysis, oxidation (Nakhaei et al., 2021), and enzymatic degradation (Mumtaz and Reimhult, 2018). The composition is essential for maintaining their chemical stability. Lipid molecules with higher degrees of unsaturation are more prone to oxidation, whereas lipids with ester bonds are susceptible to hydrolysis. Environmental factors such as temperature, pH, and existing metal ions significantly influence liposome chemical stability. Higher temperatures and acidic pH conditions can accelerate degradation processes. The incorporation of stabilising agents, like antioxidants and chelants, can improve the chemical stability of liposomes by preventing oxidation and metal-ion-induced degradation. Storage conditions, including the degree of temperature and light exposure to light, impact the chemical stability of liposomes. Low-temperature storage and protection from light can help preserve the integrity of liposomal formulations. Analytical techniques, such as spectroscopy, chromatography, and microscopy, are valuable tools for assessing the chemical stability of liposomes and monitoring changes in structure and composition over time. Understanding the stability of liposomes at the molecular level is of paramount importance in the realm of formulation development and storage, ensuring their effectiveness and safety as drug delivery systems (Grit and Crommelin, 1993).

- *In vivo* degradation *In vivo* degradation

The concentration and activity affect the release and degradation rate. Higher concentrations of PLA2 and increased enzymatic activity result in faster degradation and release of encapsulated molecules. The size and composition of the bilayer vesicle affect its susceptibility to PLA2-induced degradation. Smaller structures that self-assemble into a bilayer structure generally exhibit faster degradation and release compared to larger ones. Additionally, the presence of specific lipid components, such as phosphatidylcholine, can modulate the degradation rate (Mumtaz and Reimhult, 2018).

4.1.3. Suboptimal Bio-Distribution

According to research, the biodistribution of liposomes containing synthetic galactose-terminated diacylglyceryl-poly (ethyleneglycol)s was found to be suboptimal. When these liposomes were administered intravenously, they showed limited accumulation in the liver, which is the desired target organ for this particular formulation. Instead, the liposomes accumulated in the spleen and lungs, suggesting a suboptimal bio-distribution pattern. This indicates that further optimisation is needed to improve the targeting and distribution of these liposomes to the liver (Shimada et al., 1997). However, decades later, a team likely investigated the therapeutic effects of liposomes and exosomes in the sepsis model. This research was conducted to assess the capacity of nanocarriers to transport therapeutic agents, modulate immune responses, reduce inflammation, or promote tissue regeneration. In this study, they observed that when injected intravenously, exosomes were collected in large amounts in the lungs of sepsis mice and their presence in the bloodstream was prolonged due to liver dysfunction. However, liposomes did not show these specific effects related to sepsis. These findings provide valuable information on how exosomes and liposomes are distributed in organs, especially in the context of sepsis (Mirzaaghasi et al., 2021).

Table 11. Challenges associated with liposome-mediated drug delivery

Challenges associated with liposome formulation	
Cost	• *Encapsulation Efficiencies*: The conventional method of producing liposomes suffers from low encapsulation efficiencies and challenges in controlling the particle size distribution. This leads to the wastage of encapsulated molecules and increased production costs. Economic analysis for industrial scale-up is necessary (Trucillo et al., 2020). • *Production Cost*: Liposome manufacturing is expensive, especially on large lots. An ideal preparation method, especially for large-scale production, should involve fewer steps and equipment allowing quick and consistent synthesis with minimal equipment (Mozafari, 2005).
Sterilization	Filtration is the preferred method for sterilisation because other treatments can damage the liposomes. Filtration, however, cannot remove viruses, so strict microbiological control is needed. Filtration takes a long time, especially for large batches of liposomal products. Ordinary heat sterilisation at 121°C for 20 minutes can maintain the structure of liposomes and preserve their high encapsulation efficiency (Mozafari, 2005).
Lyophilization	Lyophilization, despite having limited examples of marketed products, remains the main technique studied for drying liposomal formulations. The complexity is in selecting the appropriate excipients and processing variables to safeguard the integrity of the liposome membrane during freezing and dehydration. Since liposomes inherently contain water and depend on their presence for assembly, their removal during drying can lead to significant and sometimes irreversible structural changes (Franzé et al., 2018).
Manufacturing Processes	The manufacturing processes currently used for liposomes have several drawbacks that hinder their development and production. These include the need for smaller particles through specialised equipment, the usage of multi-step batch procedures, and the use of small batch sizes. These challenges not only drive up costs but also restrict production capacity (Shah et al., 2020).
Large Scale Production	Large-scale liposome manufacture currently relies on the ethanol injection method followed by extrusion. However, this approach presents challenges in terms of formulation development, process scalability, and technology transfer due to its complexity and the numerous unit operations involved. Essential steps such as lipid hydration, membrane extrusion, and diafiltration are resource-intensive, demanding expertise, and strict process controls (Shah et al., 2020).

4.2. Approaches to Overcome Challenges in Liposome-Mediated Drug Delivery

Liposomal dosage forms biodistribution, effectiveness, and safety might be affected by problems including lipid oxidation and hydrolysis, drug leakage, vesicle coagulation, or fusion, which provide difficulties in their creation. Although formulation approaches and storage conditions can help mitigate these issues, drying liposome-based formulations becomes necessary when other measures are not sufficient. This is particularly the case for liposomes intended for oral or pulmonary administration (Franzé et al., 2018).

4.2.1. Approach to Protect Physical Stability

When the sterol content, the enthalpy of the gel to liquid crystalline phase is reduced (Nakhaei et al., 2021).

4.2.2. Approach to Protect Chemical Stability

To protect liposomes from oxidative degradation, several measures can be taken. These include shielding them from light, incorporating antioxidants; alpha-tocopherol or butylated hydroxytoluene (BHT), processing the product in an oxygen-free environment (using nitrogen or argon), or incorporating EDTA to remove traces of heavy metals. Lysophosphatidylcholine (lysoPC), which improves the permeability of the liposomal contents, can be produced via hydrolysis of the ester link in the glycerol moiety of phospholipids. As a result, it is critical to control quantities of lysoPC in the liposomal medicinal product. Using lysoPC-free phosphatidylcholine to create liposomes is one method to accomplish this (Sharma et al., 2018).

Although liposome stability enhancement strategies are still under investigation, excipients such as disaccharides, in particular, are incorporated into the aqueous layer of dispersions of liposomes to enhance the stability of the liposome membrane in the freeze-drying process. These sugars interact with phospholipid head groups through hydrogen bonds, bridging multiple phospholipids simultaneously. The aforementioned interaction serves to augment the separation between the polar heads of the phospholipids, thereby causing a decrease in the degree of van der Waals forces that occur amidst chains of hydrocarbons and lowering the bilayer transition temperature (Tm). The optimal sugar-to-lipid ratio for stabilisation is 5:1 and encapsulating the sugar both inside and outside the liposomes further enhances stability (Franzé et al., 2018).

4.2.3. Large-Scale Production

To overcome this challenge, alternative techniques such as nanoprecipitation /antisolvent and microfluidics show promise. When exposed to an aqueous environment, phospholipids and a stabiliser mixed in a biocompatible solvent cause the spontaneous production of submicrometer vesicles, a process known as nanoprecipitation. This self-assembled method enables the production of liposomes with specific particle sizes and low polydispersity indices, eliminating the need for extrusion or organic solvents. On the contrary, microfluidics enables fine control of liposome precipitation, resulting in desired particle sizes without requiring separate lipid hydration and extrusion steps. These innovative approaches have the potential to streamline the liposome manufacturing process, enhancing its efficiency and robustness (Shah et al., 2020).

4.2.4. In-Vivo Degradation

The investigation demonstrated the ability to employ lipid-containing polymersomes responsive to phospholipase A2 (PLA2) as vehicles for the dispensation of medicine, whereby the release of medicaments can be instigated by the existence of PLA2 enzymes in the precise biological environment (Mumtaz and Reimhult, 2017). By altering certain properties of liposomes, particularly cholesterol, one may anticipate a potential impact on blood clearance and tissue distribution of intravenously administered liposomes. Modification of liposomal cholesterol content has been observed to have an inhibitory effect on reticuloendothelial uptake (Nakhaei et al., 2021).

4.2.5. Economic Approach To Manufacturing Liposomes

Investigators conclude their research with a statement that the process bears a comparatively modest investment cost, suggesting that the liposome manufacturing process discussed, known as SuperLip, is not costly to implement (Trucillo et al., 2020). This indicates that the process is economically advantageous compared to conventional techniques. The profitability of the process is attributed to both the product quality and the lower investment cost, which allows for a positive income and a high return on investment. Furthermore, the potential for increased sales prices and production rates, as well as the consideration of scaling up the process to an industrial level, further emphasises that liposome production using SuperLip is not expensive. Advantages of SuperLip over traditional methods of liposome formation;

a) The SuperLip process achieves molecule encapsulation efficiencies of over 95%.
b) The high encapsulation efficiency of SuperLip is seen as an opportunity for economic profitability.
c) The Liposomes produced by SuperLip at the nanometric level allow for precise drug administration in small spaces within human tissues.
d) SuperLip generates monodispersed liposome samples, ensuring consistent and reproducible production.
e) The one-step production and continuous nature of SuperLip provide advantages for industrial applications, resulting in cost reduction and increased profitability compared to other processes.

4.3. Prospects of Liposome-Mediated Drug Delivery

Liposomes as drug carriers were used earlier than nanomedicines, with liposomal applications in various fields beginning in 1970 compared to 1990 for nanomedicines. The number of publications related to medicine carriers has steadily increased over time, reaching 74% in 2020. However, the percentage of drug nanoliposomes within total nanomedicine publications remains remarkably low at 7%. This discrepancy may be attributed to the less frequent use of the term "nano" in combination with "liposome" compared to other nanoparticle-related terms. The development of smart liposomal systems uses microenvironmental stimuli to trigger drug release. Although hopeful results have been attained in preclinical investigation, challenges remain in clinical translation, including drug leakage, patient variability, and complexities of multimodal therapies (Liu et al., 2022).

It is important to conduct a thorough examination of the potential acute or chronic noxiousness of advanced liposomes in both the human species and the ecosystem. Furthermore, with the increasing prevalence of nanomedicines, it is imperative to conduct thorough research on their cost-effectiveness and devise strategies to enhance their accessibility. These areas require further research and input to ensure the safe and cost-effective implementation of nanomedicine technologies (Patra et al., 2018).

Conclusion

Liposomes and lipid-based nanomedicines have shown promise in healthcare, improving drug delivery and reducing toxicity. Liposomes manifest an extensive array of applications, including but not limited to the delivery of antifungal agents, antibiotics, anti-neoplastic drugs, genetic therapeutics, and imaging agents. Complex manufacturing processes limit the widespread application and scalability of liposomes. New techniques for manufacturing must be devised to overcome these difficulties and broaden the use of liposomes in clinical settings. Efficient analytical tools are required to ensure the quality and effectiveness of liposomal drug products. These advances in manufacturing processes and analytical tools can enhance the development and application of lipid-containing liposomes and nanotherapeutics in various diseases. The development of bioconjugation techniques for the integration of drugs into liposomes and their targeted delivery have demonstrated marked advancement. Liposomes have been used successfully as RNA transport carriers in COVID-19 immunogen contenders. These liposome-based carriers have advanced to clinical trials and have received authorisation for use in humans. Ongoing clinical trials are exploring liposome formulations with bioconjugation strategies to improve drug efficacy and reduce side effects. Additional investigation would enhance the configuration of liposomes for particular uses and gain an understanding of their impacts on cells and their surrounding environments. Conventional methods face challenges in achieving optimal liposome characteristics and industrial-scale production. Challenges such as manufacturing, quality assurance, and cost must be addressed for further advancements in liposomal technology. Advanced liposome preparation technologies are being researched to improve their size, shape, and encapsulation efficiency. Supercritical fluid (SCF)-based methods offer a greener alternative and can produce high-quality liposomes. More research is needed to optimize process design and overcome challenges in industrial-scale liposome production using SCF methods. Combined methods that incorporate both SCF-assisted and conventional techniques are recommended for liposome preparation. Liposome publications related to the drug carrier have steadily increased over time, while nanomedicine publications have grown exponentially. Liposome development is progressing with approvals. The safety and affordability of advanced liposomes require further analysis and research for successful implementation.

Acknowledgments

We acknowledge BioRender.com for its valuable contribution to visually represent our scientific ideas.

References

Abud MB, Louzada RN, Isaac DL, Souza LG, Dos Reis RG, Lima EM, de Vila MP. Evaluation of *in vivo* and *in vitro* toxicity of liposome-encapsulated sirolimus. *International Journal of Retina and Vitreous*. 2019 Dec;5:1-0.

Agarwal K. Liposome-Assisted Drug Delivery-An Updated Review. Indian *Journal of Pharmaceutical Sciences*. 2022 July 14;84(4):797-811.

Agrawal M, Tripathi DK, Saraf S, Saraf S, Antimisiaris SG, Mourtas S, Hammarlund-Udenaes M, Alexander A. Recent advances in liposomes targeting strategies to cross the blood-brain barrier (BBB) for the treatment of Alzheimer's disease. *Journal of controlled release*. 2017 Aug. 28;260:61-77.

Akbarzadeh A, Rezaei-Sadabady R, Davaran S, Joo SW, Zarghami N, Hanifehpour Y, Samiei M, Kouhi M, Nejati-Koshki K. Liposome classification, preparation, and applications. *Nanoscale research letters*. 2013 Dec;8(1):1-9.

Allen, T. M., Cullis, PR. Drug delivery systems: entering the mainstream. *Science*. 2004 Mar 19;303(5665):1818-22.

Allen, T. M., Cullis, PR. Liposomal drug delivery systems: from concept to clinical applications. *Advanced drug delivery reviews*. 2013 Jan 1;65(1):36-48.

Almeida B, Nag OK, Rogers KE, Delehanty JB. Recent progress in bioconjugation strategies for liposome-mediated drug delivery. *Molecules*. 2020 December 1;25(23):5672.

Alrbyawi H, Poudel I, Annaji M, Boddu SH, Arnold RD, Tiwari AK, Babu RJ. pH-sensitive liposomes for enhanced cellular uptake and cytotoxicity of daunorubicin in melanoma cell lines. *Pharmaceutics*. 2022 May 26;14(6):1128.

Anderson M, Omri A. Effect of different lipid components on the *in vitro* stability and release kinetics of liposome formulations. *Drug delivery*. 2004 Jan 1;11(1):33-9.

Araki R, Matsuzaki T, Nakamura A, Nakatani D, Sanada S, Fu HY, Okuda K, Yamato M, Tsuchida S, Sakata Y, Minamino T. Development of a novel one-step production system for injectable liposomes under GMP. *Pharmaceutical Development and Technology*. 2018 Jul 3;23(6):602-7.

Arias-Alpizar G, Kong L, Vlieg RC, Rabe A, Papadopoulou P, Meijer MS, Bonnet S, Vogel S, van Noort J, Kros A, Campbell F. Light-triggered switching of liposome surface charge directs delivery of membrane impermeable payloads *in vivo*. *Nature Communications*. 2020 Jul 20;11(1):3638.

Balon K, Riebesehl BU, Müller BW. Determination of the liposome partitioning of ionisable drugs by titration. *Journal of pharmaceutical sciences*. 1999b Aug;88(8):802-6.

Balon K, Riebesehl BU, Müller BW. Drug liposome partitioning as a tool for the prediction of human passive intestinal absorption. *Pharmaceutical research*. 1999a Jun;16:882-8.

Bangham AD, Horne RW. Negative staining of phospholipids and their structural modification by surface-active agents as observed in the electron microscope. *Journal of molecular biology*. 1964 Jan 1;8(5):660-IN10.

Bangham AD, Standish MM, Watkins JC. Diffusion of univalent ions across the lamellae of swollen phospholipids. *Journal of molecular biology*. 1965 Aug 1;13(1):238-IN27.

Barenholz, Y. Liposome application: problems and prospects. *Current opinion in colloid & interface science*. 2001 Feb. 1;6(1):66-77.

Bian J, Girotti J, Fan Y, Levy ES, Zang N, Sethuraman V, Kou P, Zhang K, Gruenhagen J, Lin J. Fast and versatile analysis of liposome encapsulation efficiency by nanoparticle exclusion chromatography. *Journal of Chromatography A*. 2022 Jan 11;1662:462688.

Buboltz JT, Feigenson GW. A novel strategy for the preparation of liposomes: rapid solvent exchange. *Biochimica et Biophysica Acta (BBA) biomembranes*. 1999 Mar 4;1417(2):232-45.

Chen, R., Li, R., Liu, Q., Bai, C., Qin, B., Ma, Y., Han, J. Ultradeformable liposomes: a novel vesicular carrier for enhanced transdermal delivery of procyanidins: effect of surfactants on the formation, stability, and transdermal delivery. *AAPS PharmSciTech*. 2017 Jul;18:1823-32.

Chu, CJ, Szoka, FC. pH-Sensitive Liposomes. *Journal of Liposome Research.* 1994; 4(1):361–95.

Costa AP, Xu X, Khan MA, Burgess DJ. Liposome formation using a coaxial turbulent jet in co-flow. *Pharmaceutical research.* 2016 Feb;33:404-16.

Crommelin, DJ, van Hoogevest, P, Storm, G. The role of liposomes in the development of clinical nanomedicine. What now? Now what? *Journal of Controlled Release.* 2020 Feb 1; 318:256-63.

Düzgüneş N, Nir S. Mechanisms and kinetics of liposome–cell interactions. *Advanced drug delivery reviews*. 1999 Nov 10;40(1-2):3-18.

Franzé S, Selmin F, Samaritani E, Minghetti P, Cilurzo F. Lyophilization of liposomal formulations: still necessary, still challenging. *Pharmaceutics*. 2018 Aug 28;10(3):139.

Gaber MH, Wu NZ, Hong K, Huang SK, Dewhirst MW, Papahadjopoulos D. Thermosensitive liposomes: extravasation and release of contents in tumour microvascular networks. *International Journal of Radiation Oncology* Biology* Physics*. 1996 Dec 1;36(5):1177-87.

Gala, RP, Khan, I, Elhissi, AM, Alhnan, MA. A comprehensive production method of self-cryoprotected nanoliposome powders. *International Journal of Pharmaceutics*. 2015 May 30;486(1-2):153-8.

Gilbert, BE, Seryshev, A, Knight, V, Brayton, C. 9-nitrocamptothecin liposome aerosol: lack of subacute toxicity in dogs. *Inhalation Toxicology*. 2002 Jan 1; 14(2):185-97.

Grit M, Crommelin DJ. Chemical stability of liposomes: implications for their physical stability. *Chemistry and physics of lipids*. 1993 Sep 1;64(1-3):3-18.

Immordino ML, Dosio F, Cattel L. Stealth liposomes: review of the basic science, rationale, and clinical applications, existing and potential. *International Journal of nanomedicine*. 2006;1(3):297.

Inglut CT, Sorrin AJ, Kuruppu T, Vig S, Cicalo J, Ahmad H, Huang HC. Immunological and toxicological considerations for the design of liposomes. Nanomaterials. 2020 Jan 22;10(2):190.

Jahn A, *Vreeland* WN, DeVoe DL, Locascio LE, Gaitan M. Microfluidic-directed formation of liposomes of controlled size. *Langmuir*. 2007 May 22;23(11):6289-93.

Johnston MJ, Edwards K, Karlsson G, Cullis PR. Influence of drug-to-lipid ratio on drug release properties and liposome integrity in liposomal doxorubicin formulations. *Journal of liposome research. 2008* Jan 1;18(2):145-57.

Kaddah S, Khreich N, Kaddah F, Charcosset C, Greige-Gerges H. Cholesterol modulates the fluidity and permeability for a hydrophilic molecule. *Food and Chemical Toxicology*. 2018 Mar 1;113:40-8.

Kang KC, Lee DH, Wang SH, Lee CI, Pyo HB, Jeong NH. Development of a nanoliposome with unsaturated lecithin. *Applied Chem*. 2004;8:1-4.

Karaz S, Senses E. Liposomes Under Shear: Structure, Dynamics, and Drug Delivery Applications. *Advanced nanobiomed research*. 2023 Feb 5;3(4):202200101.

Khan AA, Allemailem KS, Almatroodi SA, Almatroudi A, Rahmani AH. Recent strategies towards the surface modification of liposomes: An innovative approach for different clinical applications. *3 Biotech*. 2020 Apr;10:1-5.

Khan DR, Rezler EM, Lauer Fields J, Fields GB. Effects of drug hydrophobicity on liposomal stability. *Chemical Biology & Drug Design.* 2008 Jan;71(1):3-7.

Knudsen KB, Northeved H, Ek PK, Permin A, Gjetting T, Andresen TL, Larsen S, Wegener KM, Lykkesfeldt J, Jantzen K, Loft S. *In vivo* toxicity of cationic micelles and liposomes. *Nanomedicine: Nanotechnology, biology, and medicine*. 2015 Feb. 1;11(2):467-77.

Krasnici S, Werner A, Eichhorn ME, Schmitt-Sody M, Pahernik SA, Sauer B, Schulze B, Teifel M, Michaelis U, Naujoks K, Dellian M. Effect of surface charge of liposomes on their uptake by angiogenic tumour vessels. *International journal of cancer*. 2003 Jul 1;105(4):561-7.

Lee, MK. Liposomes for enhanced bioavailability of water-insoluble drugs: *in vivo* evidence and recent approaches. *Pharmaceutics*. 2020 Mar 13;12(3):264.

Lesoin L, Crampon C, Boutin O, Badens E. Development of a continuous dense gas process for the production of liposomes. *The Journal of Supercritical Fluids*. 2011 Dec 1;60:51-62.

Li J, Wang X, Zhang T, Wang C, Huang Z, Luo X, Deng Y. A review of phospholipids and their main applications in drug delivery systems. *Asian Journal of pharmaceutical sciences*. 2015 Apr. 1;10(2):81-98.

Li, SD, Huang, L. Stealth nanoparticles: High-density but sheddable PEG is a key for tumour targeting. *Journal of controlled release: the official journal of the Controlled Release Society.* 2010 Aug 8;145(3):178.

Liu P, Chen G, Zhang J. A review of liposomes as a drug delivery system: current status of approved products, regulatory environments, and future perspectives. *Molecules*. 2022 Feb. 17; 27(4):1372.

Liu X, Huang G. Formation strategies, mechanism of intracellular delivery, and potential clinical applications of pH-sensitive liposomes. *Asian Journal of pharmaceutical sciences*. 2013 Dec 1;8(6):319-28.

Lombardo D, Kiselev MA. Methods of liposome preparation: formation and control factors of versatile nanocarriers for biomedical and nanomedicine application. *Pharmaceutics*. 2022 Feb 28;14(3):543.

Maja L, eljko K, Mateja P. Sustainable technologies for liposome preparation. *The Journal of Supercritical Fluids* 2020 Nov. 1;165:104984.

Maniyar MG, Kokare CR. Formulation and evaluation of lopinavir spray dried liposomes for topical application. *Journal of Pharmaceutical Investigation*. 2019 Mar 15;49:259-70.

Maritim S, Boulas P, Lin Y. Comprehensive analysis of liposome formulation parameters and their influence on encapsulation, stability, and drug release in glibenclamide liposomes. *International Journal of pharmaceutics*. 2021 Jan 5;592:120051.

Marqués-Gallego, P., de Kroon, AI. Ligation strategies for targeting liposomal nanocarriers. *BioMed research international*. 2014 Jul 14;2014.

Miranda D, Lovell JF. Mechanisms of light-induced liposome permeabilization. *Bioengineering & translational medicine*. 2016 Sep;1(3):267-76.

Mirzaaghasi A, Han Y, Ahn SH, Choi C, Park JH. Biodistribution and pharmacokinectics of liposomes and exosomes in a mouse model of sepsis. *Pharmaceutics*. 2021 Mar 22;13(3):427.

Moghimi SM. Liposomes. In: Bhushan B, editor. *Encyclopaedia of Nanotechnology*. Dordrecht: Springer The Netherlands; 2012. p. 1218–23.

Mozafari MR. Liposomes: an overview of manufacturing techniques. *Cellular and molecular biology letters*. 2005 Jan. 1;10(4):711.

Mumtaz Virk M, Reimhult E. Phospholipase A2-induced degradation and release from lipid-containing polymersomes. *Langmuir*. 2018 Jan 9;34(1):395-405.

Nakhaei P, Margiana R, Bokov DO, Abdelbasset WK, Jadidi Kouhbanani MA, Varma RS, Marofi F, Jarahian M, Beheshtkhoo N. Liposomes: structure, biomedical applications, and stability parameters with emphasis on cholesterol. *Frontiers in Bioengineering and Biotechnology*. 2021:748.

Nsairat H, Khater D, Sayed U, Odeh F, Al Bawab A, Alshaer W. Liposomes: structure, composition, types, and clinical applications. *Heliyon*. 2022 May 1;8(5):e09394.

Ong SG, Chitneni M, Lee KS, Ming LC, Yuen KH. Evaluation of the extrusion technique for nanosizing liposomes. *Pharmaceutics*. 2016 Dec 21;8(4):36.

Ong SG, Ming LC, Lee KS, Yuen KH. Influence of liposome encapsulation efficiency and size on the oral bioavailability of griseofulvin-loaded liposomes. *Pharmaceutics*. 2016 Aug 26;8(3):25..

Patel GB, Sprott GD. Archaeobacterial ether lipid liposomes (archaeosomes) as novel vaccine and drug delivery systems. *Critical reviews in biotechnology*. 1999 Jan 1;19(4):317-57.

Patil YP, Jadhav S. Novel methods for liposome preparation. *Chemistry and physics of lipids*. 2014 Jan 1;177:8-18.

Patra JK, Das G, Fraceto LF, Campos EV, Rodriguez-Torres MD, Acosta-Torres LS, Diaz-Torres LA, Grillo R, Swamy MK, Sharma S, Habtemariam S. Nano based drug delivery systems: recent developments and future prospects. *Journal of Nanobiot*echnology. 2018 Dec;16(1):1-33.

Pidgeon C, Hunt CA. Light-sensitive liposomes. *Photochemistry and photobiology*. 1983 May;37(5):491-4.

Raju A, Muthu MS, Feng SS. TPGS liposomes conjugated with trastuzumab for sustained and targeted docetaxel delivery. *Expert opinion on drug delivery.* 2013 Jun 1;10(6):747-60.

Rushmi ZT, Akter N, Mow RJ, Afroz M, Kazi M, de Matas M, Rahman M, Shariare MH. The impact of formulation attributes and process parameters on black seed oil-loaded liposomes and their performance in animal models of analgesia. *Saudi pharmaceutical journal*. 2017 Mar 1;25(3):404-12.

Sasaki K, Kogure K, Chaki S, Nakamura Y, Moriguchi R, Hamada H, Danev R, Nagayama K, Futaki S, Harashima H. An artificial virus-like nano carrier system: enhanced endosomal escape of nanoparticles via synergistic action of pH-sensitive fusogenic peptide derivatives. *Analytical and bioanalytical chemistry.* 2008 Aug;391:2717-27.

Sawant RR, Torchilin VP. Challenges in the development of targeted liposomal therapeutics. *The AAPS journal*. 2012 Jun;14:303-15.

Sercombe L, Veerati T, Moheimani F, Wu SY, Sood AK, Hua S. Advances and challenges of liposome-assisted drug delivery. *Frontiers in Pharmacology.* 2015 Dec 1; 6:286.

Shah S, Dhawan V, Holm R, Nagarsenker MS, Perrie Y. Liposomes: Advances and innovation in the manufacturing process. *Advanced Drug Delivery Reviews*. 2020 Jan. 1;154:102-22.

Shailesh S, Neelam S, Sandeep K, Gupta GD. *Liposomes: a review Journal of Pharmacy Research*. 2009 Jul;2(7):1163-7.

Shaker S, Gardouh AR, Ghorab MM. Factors affecting the particle size prepared by ethanol injection method. *Research in pharmaceutical sciences*. 2017 Oct;12(5):346.

Sharma D, Ali AA, Trivedi LR. An Updated Review on: Liposomes as a drug delivery system. *PharmaTutor*. 2018 Feb. 1;6(2):50-62.

Shashi K, Satinder K, Parashar B. A complete review on: Liposomes. *International Journal of Pharmacy*. 2012 Jul; 3.

Shimada K, Kamps JA, Regts J, Ikeda K, Shiozawa T, Hirota S, Scherphof GL. Biodistribution of liposomes containing synthetic galactose-terminated diacylglyceryl poly (ethyleneglycol) s. *Biochimica et Biophysica Acta (BBA) biomembranes*. 1997 Jun 12;1326(2):329-41.

Streck S, Hong L, Boyd BJ, McDowell A. Microfluidics for the production of nanomedicines: Considerations for polymer- and lipid-based systems. *Pharmaceutical nanotechnology*. 2019 Dec 1;7(6):423-43.

Subramaniam R, Xiao Y, Li Y, Qian SY, Sun W, Mallik S. Light-mediated and H-bond facilitated liposomal release: the role of lipid head groups in release efficiency. *Tetrahedron Letters*. 2010 Jan 20;51(3):529-32.

Suk JS, Xu Q, Kim N, Hanes J, Ensign LM. PEGylation as a strategy for improving nanoparticle-based drug and gene delivery. *Advanced drug delivery reviews*. 2016 Apr 1;99:28-51.

Sun D, Lu ZR. Structure and Function of Citric and Ionizable Lipids for Nucleic Acid Delivery. *Pharmaceutical Research.* 2023 Jan 4:1-20.

Sur S, Fries AC, Kinzler KW, Zhou S, Vogelstein B. Remote loading of preencapsulated drugs into stealth liposomes. *Proceedings of the National Academy of Sciences.* 2014 Feb 11;111(6):2283-8.

Ta, T., Porter, TM. Thermosensitive liposomes for localised delivery and triggered release of chemotherapy. *Journal of controlled release.* 2013 Jul 10;169(1-2):112-25.

Teixeira MC, Carbone C, Souto EB. Beyond liposomes: Recent advances in lipid-based nanostructures for poorly soluble/poorly permeable drug delivery. *Progress in lipid research.* 2017 Oct 1;68:1-1.

Tiwari G, Tiwari R, Sriwastawa B, Bhati L, Pandey S, Pandey P, Bannerjee SK. Drug delivery systems: An updated review. *International journal of pharmaceutical investigation.* 2012 Jan;2(1):2.

Torchilin, VP. Recent advances with liposomes as pharmaceutical carriers. *Nature reviews drug discovery.* 2005 Feb 1;4(2):145-60.

Trucillo P, Campardelli R, Reverchon E. Antioxidant-loaded emulsions entrapped in liposomes produced using a supercritical assisted technique. *The Journal of Supercritical Fluids* Dec. 1;154:104626.

Trucillo P, Campardelli R, Reverchon E. Supercritical CO2-assisted liposome formation: Optimisation of the lipid layer for an efficient hydrophilic drug loading. *Journal of Co2 utilisation.* 2017 Mar 1;18:181-8.

Trucillo, P., Campardelli, R., Iuorio, S., De Stefanis, P., Reverchon, E. Economic analysis of a new business for liposome manufacturing using a high pressure system. *Processes.* 2020 Dec 6;8(12):1604.

Urits I, Swanson D, Swett MC, Patel A, Berardino K, Amgalan A, Berger AA, Kassem H, Kaye AD, Viswanath O. A review of patisiran (ONPATTRO®) for the treatment of polyneuropathy in people with hereditary transthyretin amyloidosis. *Neurology and therapy.* 2020 Dec;9:301-15.

Vakili-Ghartavol R, Rezayat SM, Faridi-Majidi R, Sadri K, Jaafari MR. Optimisation of Docetaxel loading conditions in liposomes: Proposal of potential products for metastatic breast carcinoma chemotherapy. *Scientific reports.* 2020 Mar 27;10(1):5569.

Veloso DF, Benedetti NI, vila RI, Bastos TS, Silva TC, Silva MR, Batista AC, Valadares MC, Lima EM. Intravenous delivery of a liposomal formulation of voriconazole improves drug pharmacokinetics, tissue distribution, and enhances antifungal activity. *Drug delivery.* 2018 Jan. 1;25(1):1585-94.

Wang SJ, Huang WS, Chuang CM, Chang CH, Lee TW, Ting G, Chen MH, Chang PM, Chao TC, Teng HW, Chao Y. A phase 0 study of the pharmacokinetics, biodistribution, and dosimetry of 188 Re-liposome in patients with metastatic tumours. *EJNMMI research.* 2019 Dec;9:1-3.

Webb C, Khadke S, Tandrup Schmidt S, Roces CB, Forbes N, Berrie G, Perrie Y. Impact of solvent selection: strategies to guide the manufacturing of liposomes using microfluidics. *Pharmaceutics.* 2019 Dec 4;11(12):653.

Xia Y, Sun J, Liang D. Aggregation, fusion, and leakage of liposomes induced by peptides. *Langmuir.* 2014 Jul 1;30(25):7334-42.

Zhang, Y., Cao, Y., Luo, S., Mukerabigwi, JF, and Liu, M. Nanoparticles as drug delivery systems of combination therapy for cancer. *In Nanobiomaterials in Cancer Therapy* 2016 Jan 1 (pp. 253-280). William Andrew Publishing.

Zolnik BS, González-Fernández, Sadrieh N, Dobrovolskaia MA. Minireview: Nanoparticles and the immune system. *Endocrinology*. 2010 Feb 1;151(2):458-65.

Chapter 3

Liposomes in Cancer Therapy: Current State and Future Directions

Asad Ahmad[1]
Aditya Singh[1]
Shubhrat Maheshwari[2]
and Anas Islam[1,*]
[1]Faculty of Pharmacy, Integral University, Lucknow, Uttar Pradesh, India
[2]Faculty of Pharmaceutical Sciences, Rama University Mandhana, Bithoor Road, Kanpur, Uttar Pradesh, India

Abstract

Liposomes are nanosized vesicles composed of phospholipid bilayers that can encapsulate various drugs and deliver them to cancer cells and tissues. The unique composition of liposomes enables encapsulation of both hydrophilic and hydrophobic drugs, resulting in improved pharmacokinetics, reduced toxicity, and increased efficacy compared to free drugs. Liposomes offer several advantages over conventional drug delivery systems, such as enhanced solubility, stability, biocompatibility, and specificity. In cancer therapy, liposomes offer several advantages, including targeted drug delivery, enhanced permeability and retention effect, and controlled drug release. This chapter provides a comprehensive overview of the current state and future directions of liposome-based cancer therapy. It covers the strategies for achieving active and passive targeting of liposomes to cancer sites, the applications of liposome-based drugs for different types of cancer therapy, and the emerging trends and opportunities for

* Corresponding Author's Email: anasislam051@gmail.com.

In: Liposomes
Editors: Usama Ahmad and Anas Islam
ISBN: 979-8-89113-636-6

improving the efficacy and safety of liposome-based cancer therapy. The chapter also discusses the potential barriers and solutions for translating liposome-based cancer therapy from laboratory to clinic. We review the current applications of liposomes in cancer therapy, including FDA- approved liposome-based drugs and clinical trials. Furthermore, we highlight the challenges and limitations of liposome-based therapies. We conclude that liposomes hold great promise in cancer therapy, and continued research and development will further enhance their potential to improve patient outcomes.

Keywords: liposomes, cancer therapy, FDA, targeting, enhanced permeability

1. Introduction

Among other complicated and hazardous illnesses that are still largely incurable, cancer is one of the supreme origins of mortality. Nonetheless, there has been significant advancement in this field. To treat cancer and its effects, a number of techniques have been devised; comprise hyperthermia, phototherapy, gas therapy, chemotherapy, and radiotherapy (Cho HJ et al., 2020). Additionally, because conventional chemotherapy frequently fails to distinguish between cancerous and healthy cells, it is unable to treat cancer in a targeted manner. Most of the medications employed in this treatment have undesirable side effects instead of affecting the malignant tissue that is their intended target (Garcia MA et al., 2020). Controlled drug delivery systems are therefore strongly advised (Arévalo-Pérez R et al., 2020). These new techniques add to existing ones, are more precise and efficient, and solely identify and target tumor cells. Liposomes is used for formation of controlled and targeted drug delivery system to targeting the cancerous cells (Crivelli B et al., 2018). However, the adoption of liposomes has been slow due to their classification as "novel" by regulators, even when prepared using compendial excipients (Zhu YS et al., 2021). Liposomes can help overcome formulation challenges by enhancing the performance and processability of resulting materials (Mouhid L et al., 2017). "By reducing manufacturing complexity and, as a result, drug development time, Liposomes can significantly expedite time-to-market," he adds (Aziz A et al., 2022). Liposomes analysis includes clinical indications for gene treatments, metabolic and infectious disorders, cancer, dental and ophthalmic conditions, and other conditions (Radmoghaddam ZA et al., 2022). Parenteral and oral formulations, including tablets, capsules, liquids, and injectables, as well as topical applications, nutraceuticals, over-the-counter products, and other uses, are among the various applications of pharmaceutical excipients (Luiz MT et al., 2022). The use of liposomes is aimed at enhancing the functionality of the pharmacological product, according to Gomes (Amin M et al., 2022). An increasing number of new APIs have been discovered due to advancements in high-throughput screening technology. However, approximately 75% of novel drug candidates suffer from poor aqueous solubility and inadequate bioavailability, which can be attributed to the growing structural complexity of therapeutic prospects (Khan MI et al., 2022). To overcome this challenge, various tried-and-tested as well as cutting-edge approaches, such as cyclodextrin inclusion, microemulsion, nanocrystals, cocrystals, and amorphous dispersions, are utilized to enhance the delivery of class IV pharmaceuticals compounds (Bigham A et al., 2022). Among these approaches, amorphization of pharmaceuticals has emerged as one of the most successful methods for increasing solubility and dissolution, thereby improving therapeutic bioavailability. Amorphous solids have higher internal energy than their crystalline counterparts and lack long-range order in molecular packing (Van NH et al., 2022). To stabilize amorphous materials and improve their oral bioavailability, polymer-based amorphous solid dispersions have been widely employed. However, this approach also has its limitations, as noted in various reviews (Rahman M et al., 2021). Additionally, high-dose medication formulations may encounter issues due to the larger end product volume. Despite these limitations, pharmaceutical companies still prefer direct compression (DC) as the most popular method (Kurakula M et al., 2021). This machinery process involves a straightforward physical mixing of the active medicinal in liposomes. Liposomes used in the DC process must perform various functions, such as facilitating acceptable excellent binding capacity, to produce tablets effectively. However, finding a single material with all these desirable attributes is a challenging task. Chemotherapeutic drugs as shown in Table 1, have a limited therapeutic window and a high potential for toxicity (Aanisah N et al., 2022). Consequently, reducing chemotherapy's negative effects by only administering it to tumor locations would boost their effectiveness and improve patient care. The subject of drug distribution has undergone a revolution thanks to nanotechnology (Tranová T et al., 2022). The creation of several kinds of nanoparticles lie liposomes with sizes ranging from 10 to 1000 nm has enhanced the transport of numerous pharmacological molecules, particularly chemotherapeutic drugs, and offered creative, alternative solutions to many of the problems relating to their efficiency and safety (Tranová T et al., 2022). Liposomes are ideal medication carriers due to a variety of unique characteristics they have (Abdelhamid M et al., 2022). The main advantage of liposomes systems is that they can improve the solubility of poorly water-soluble drugs, thereby enhancing their absorption and distribution in the body (Gao Y et al., 2022). Targeting needs a molecular recognition method used to achieve "active targeting," another technique (Zupančič O et al., 2022).

Table 1. FDA approved drugs used in the treatment of cancer

Drug	Target	Cancer Stage	Date of Approval
Atezolizumab	PD-L1 expression	Metastatic NSCLC	05/18/2020
Brigatinib	Anaplastic lymphoma kinase	Adult patients with Anaplastic lymphoma kinase (ALK)-positive metastatic NSCLC	05/22/2020
Sotorasib	RAS GTPase family inhibitor, for adult patients with KRAS G12C mutated	Locally advanced or metastatic NSCLC	05/28/2021
Capmatinib	Tumors had a mutation that caused exon 14 of the mesenchymal-epithelial transition (MET) to be skipped.	Metastatic NSCLC	05/06/2020
Cemiplimab	Tumors with PD-L1 expression	Advanced NSCLC	02/22/2021
Lorlatinib	ALK-positive	Metastatic NSCLC	03/03/2021

2. Liposomes in Cancer Therapy

The word "liposome" refers to the phospholipids that make up its structural elements rather than its size, and it may be created in a range of sizes utilizing either unilamellar or multilamellar assembly (Nakmode D et al., 2022). The material used to create a cell membrane is used to create a microscopic bubble (vesicle). Utilizing liposomes that have been drug-loaded, medications for cancer and other conditions can be delivered. The basic constituents of membranes are usually phospholipids as in Table 2, which are molecules with head and tail groups. Water repels the hydrocarbon-based tail, which is made up of lengthy chains, but the hydrocarbon-based head is drawn to it. Phospholipids are present in stable, two-layer membranes (Dhaval M et al., 2022). The heads are drawn to water when it is present and form a surface that faces the water (Lazić I et al., 2022). The tails align to create a surface distance from the water since water repels them. The outer layer of the cell is exposed as one layer of the head is drawn to the nearby body of water (Singh A et al., 2021). The water within the cell has drawn a second layer of heads that are facing the inside. When multiple concentric bilayers are present, they are referred to as large multilamellar vesicles. On the other hand, a single bilayer encapsulating the aqueous core is described as a small or large unilamellar vesicle (Singh A et al., 2020). Micelles are referred to as monolayer layers, whereas liposomes are referred to as bilayer layers (Gao Y et al., 2022). Biologically active substances can be easily delivered via liposomes, which are globular lipid bilayers with a diameter of 50–1000 nm. The topical application of liposomes in dermatology and their utilization for anticancer treatment holds significant promise in minimizing the adverse effects associated with standalone pharmaceutical delivery. Additionally, they have the potential to prolong the circulation time of medications, thereby enhancing their effectiveness (Liu P et al., 2022). By affixing amino acid fragments from proteins or antibodies or other relevant pieces that target certain receptor sites, liposomes can be used to target exact cells (Chavda VP et al., 2022). Future uses for liposomes include DNA vaccination and enhanced gene therapy effectiveness, to name a few (Moosavian SA et al., 2022).

3. Different Pathways to Targeted Site for the Management of Cancer

In the targeted distribution of formulation for the treatment of cancer, liposomes have a specific function (Mohammadi M et al., 2023). The unique characteristics of the tumor microenvironment can be exploited for passive targeting of drugs, while ligands or antibodies can be used for active targeting of tumor cells. Signal transduction pathways play a critical role in the targeted delivery of drugs for cancer management using API These pathways are the series of chemical reactions that occur inside cells in response to extracellular signals, leading to changes in gene expression, protein activity, and cellular behavior (Chen M et al., 2023). Liposomes have the capability to utilize distinct signaling pathways and leverage the specific attributes of the tumor microenvironment for precise drug targeting in cancer therapy (Akbari J et al., 2022). These strategies offer the potential to enhance drug efficacy and reduce toxicity, providing a promising approach for the development of more effective and less harmful cancer therapies. There are different pathways through which can target the tumor site for cancer management, and some of these pathways are discussed below (Nguyen TT et al., 2022).

Table 2. Excipients used in the formulation of liposomes system

Processed Excipients	Applications
Glyceryl behenate (CompritolÒ 888 ATO)	A lubricant or modified release agent found in tablets that has a high melting point. is also applicable to lipid coating techniques and serves as a lipid carrier for nanoparticles.
Glyceryl monostearate (ImwitorÒ900)	The glycerol ester of stearic acid is glycerol monostearate, also referred to as GMS. It is frequently utilised in food as an emulsifier.
Soy lecithin (LipoidÒ S 75, LipoidÒ S 100)	Purified soybean phospholipids, which begin as crude soybean lecithin, serve a variety of purposes in medical, cosmetic, and food applications. As wetting agents, solubilizers, emulsifiers, liposome builders, and technical aids, these carefully specified products exhibit remarkable performance. In addition, they provide choline and vital fatty acids. Most goods are accessible in non-GMO varieties.
Egg lecithin	The highly purified portions that come from hen egg yolks are called egg phospholipids. Egg phospholipids are ideal for parenteral applications because they closely resemble the makeup of human cells. They are primarily found in fat emulsions used in drug-containing formulations or for parenteral feeding.
Poloxamer 188	A nonionic block linear copolymer known as Poloxamer 188 (P188) has cytoprotective, antithrombotic. Once licenced by the FDA in the 1960s as a therapeutic agent to lessen blood viscosity prior to transfusions, P188 is no longer contained in any products that have received Regulatory approval. Owing to its sufactant qualities, P188 is also present in over-the-counter (OTC) items like toothpaste, bowel cleaners, and mouthwash and is employed in a variety of cosmetic, business, and pharmaceutical applications.
Phosphatidylcholine (EpikuronÒÒ170, Epikuron 200)	Wetting and dispersion agent, suspension and emulsifying agent and dispersion agent, anti-crystallization agent, for tablets and powders

3.1. AGEs RAGEs Pathways

The overexpression of the receptor for advanced glycation end products (RAGE) in many types of cancer cells has been nearer to cancer development and progression. RAGE activation can promote cell proliferation, migration, invasion, and angiogenesis, which are essential for tumor growth and metastasis. Moreover, RAGE can activate signaling passage such as PI3K/AKT, MAPK/ERK, and NF-κB, which can promote tumor growth and survival and contribute to resistance to chemotherapy and radiation therapy (Boztepe T et al., 2023). Additionally, RAGE activation has been shown to regulate the tumor microenvironment by promoting the release of cytokines, chemokines, and growth factors that recruit immune cells, promotes inflammation, angiogenesis, and tissue remodeling, ultimately creating a favorable environment for tumor growth and metastasis. Preclinical investigations targeting RAGE using small molecule inhibitors, antibodies, or RAGE decoy receptors have showed encouraging outcomes, and clinical trials are now being conducted to assess their efficacy in the therapy of cancer (Li Z et al., 2023).

3.2. Transforming Growth Factor-Beta (TGF-β) Pathway

The TGF- pathway with liposomes has significant implications in the emergence of liver cancer, manifesting in two distinct ways. In the initial stages, TGF- triggers apoptosis and limits cell development, thereby preventing cancer progression. However, in response to this cytokine, cells may cause cell migration and invasion after they become resistant to the suppressive effects of TGF-. Notably, in HCC cells, heightened autocrine TGF- expression is associated with a mesenchymal, migratory, and invasive character. While not all cells undergo a complete EMT, this process remains pivotal in the advancement of liver cancer (Gugleva et al., 2023).

3.3. Epidermal Growth Factor Receptor (EGFR) Pathway

The discovery and clinical application of EGFR-targeting inhibitors hold substantial promise for augmenting the effectiveness of emerging lung cancer treatments utilizing liposomal molecules. This extends to the burgeoning domain of targeted cancer therapeutics employing liposomes (Wasim et al., 2022). The effective application of tailored therapeutic interventions in common epithelial malignancies relies on a deep understanding of the genetic variability within epithelial tumors and the development of strategies to hinder their rapid acquisition of resistance to targeted kinase inhibitors (Maheshwari S et al., 2023). The EGFR family in humans is composed of four closely related receptors that are glycoproteins with transmembrane properties. These receptors possess an internal tyrosine kinase domain as well as an external domain for binding to ligands. The activation of EGFR receptors triggers various signaling pathways such as

PI3 kinase, Ras-Raf-MAPK, JNK, and PLC, all of which have a wide array of biological effects. At the cellular level, the ligands alter adhesion and motility, inhibit apoptosis, and augment cell proliferation (Singh A et al., 2022). At the physiological level, ligands provoke the stimulation of blood vessel formation and invasion. In a three-dimensional context, activation of EGFR family members promotes the invasion and scattering of breast epithelial cells, resulting in the loss of cellular polarization and other epithelial differentiation characteristics. The EGFR/ErbB family, encompassing EGFR, HER2, ErbB3, and ErbB4, stands as the primary molecular targets for cancer treatment in current times. In the realm of breast cancer, the major and well-established therapeutic focus is on HER2, overexpressed in approximately 20-25% of cases.

3.4. Wnt/β-Catenin Signaling Pathway

The WNT/β-catenin signaling pathway is a meticulously regulated and evolutionarily conserved molecular system that governs embryonic development, cellular proliferation, and differentiation, in conjunction with liposomal interactions. Increasing evidence indicates that abnormal WNT/β-catenin signaling contributes to the onset and/or progression of liver cancer, particularly hepatocellular carcinoma (HCC) and cholangiocarcinoma (CCA), the two primary liver cancers that most frequently affect adults. The terms “non-canonical” and “canonical” refer to two distinct WNT signaling pathways, the latter of which involves the activation of β-catenin (Singh A et al., 2023).

4. Strategies for Achieving Targeted Delivery of Liposomes to Cancer Sites

The traditional method of treating cancer leads to a limited buildup of anticancer drugs at the necessary tumor site, causing unintended effects elsewhere in the body. As a result, numerous approaches have been developed and employed to specifically target and deliver anticancer drugs to the appropriate location to achieve the most optimal response in cancer therapy, utilizing liposomes. The primary strategies for delivering anticancer drugs to the tumor site are passive targeting (via the enhanced-permeability and retention effect) and active targeting (Riaz et al., 2018). Various strategies for achieving targeted delivery of liposomes to cancer sites are illustrated in Figure 1.

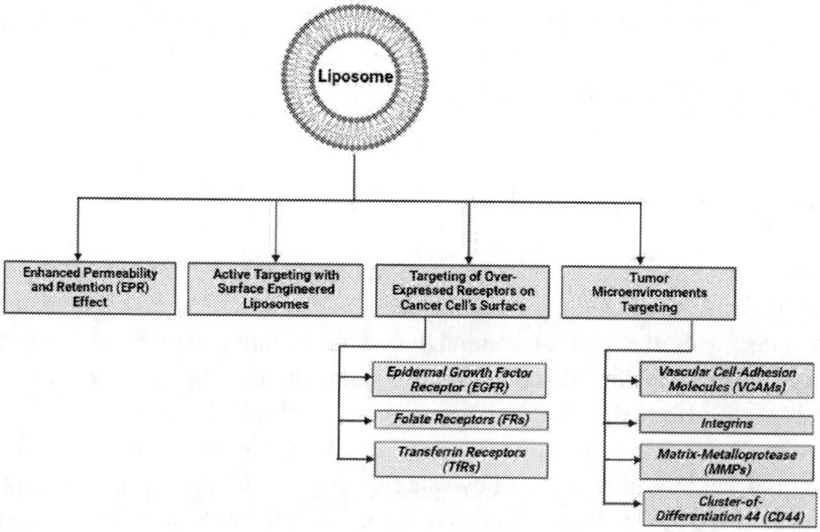

Figure 1. Different strategies for achieving targeted delivery of liposomes to cancer sites.

4.1. Enhanced Permeability and Retention (EPR) Effect and Its Application in Tumor Therapy

The reticuloendothelial system (RES) is prominently associated with traditional methods for delivering cancer therapy. The creation of a distribution system or nanocarrier with the capacity to evade destruction by the body's immune defenses, such as phagocytosis, is referred to as passive targeting. One methodology to extend the duration of circulation entails the formation of PEGylated liposomes, a technique commonly referred to as passive targeting. In addition to having increased permeability or leakiness, cancerous cells also have blocked lymphatic fluid outflow, which causes macromolecules to build up. Also known as the increased permeability and retention effect, this targeted technique (Zylberberg and Matosevic, 2016; Barenholz, 2012).

The optimization of the EPR effect can be achieved by giving thorough thought to the liposome's size. Liposomes that fall within the size range of 40 to 200 nm have exhibited greater extravasation. In recent times, proposals have been put forth to enhance the EPR effect when delivering anticancer agents to tumor locations. These proposals primarily revolve around utilizing a stimulus, whether it be internal or external, to heighten the permeability of cancer cells (Maeda et al., 2013).

4.2. Active Targeting with Surface Engineered Liposomes

The existing approach involves the guidance of the anticancer agents, specifically medications, towards the affected tissue site, namely tumor tissues, with the aim of minimizing their accumulation in non-targeted tissues or reducing off-target effects (Lim et al., 2012). This particular approach, often called active targeting or ligand-based targeting, made it possible to deliver the payload precisely to the intended location. When introduced into mice affected by cancer, the specific ligand employed in this method accurately homes in on and guides itself towards the receptors on the afflicted cells. These tailor-made liposomal structures are commonly denoted as targeted liposomes or ligand-directed liposomes (Torchilin, 2010).

Fine-tuning the ligand concentration on the surface of liposomes through suitable surface engineering methodologies is vital in developing a distinct liposomal system. The augmentation of ligand density beyond an optimal threshold may lead to the occurrence of issues such as aggregation (Feng and Mumper, 2013). Moreover, employing targeting ligands leads to improved internalization of liposomes into cancer cells (Benhabbour et al., 2012). Active targeting also possesses a distinct benefit, namely, the reduced dispersion to non-desirable tissues (Fath and Oyelere, 2016).

4.3. Targeting Overexpressed Receptors on the Surface of Cancer Cells

Contrary to normal cells, a number of receptors show elevated expression in cancer cells. Targeting these overexpressed receptors is an important technique that forms the basis for active targeting, which enhances the absorption and accretion of anticancer medicines into cancer cells at the tumor location.

4.3.1. Targeting of Epidermal Growth Factor Receptor (EGFR)

Mamot and colleagues formulated liposomes that contained doxorubicin. These liposomes consisted of DSPE, cholesterol, and MPEG-DSPE FabV fragments that were derived from cetuximab. The integration of these fragments was achieved by co-incubating them with preformed liposomes at a temperature of 55°C for a duration of 30 minutes. These immunoliposomes, which have a specific affinity for the epidermal growth factor receptor (EGFR), are specifically designed to target EGFR. The results demonstrated an increased uptake of these immunoliposomes by malignant cells and a regression in a model of human breast cancer, as compared to non-targeted liposomes (Mamot et al., 2005).

Liposomes enclosing doxorubicin, with an emphasis on the amplified presence of HER2 cancer cells, exhibited superior efficiency in drug transport and anti-cancer effects compared to untargeted liposomes. These liposomal compositions exhibited significant therapeutic outcomes across various tumor xenograft models displaying HER2 overexpression. Trastuzumab fragments served as the ligand for precise targeting of the HER2 receptor. As part of the liposome development process, Fab or scFv were covalently conjugated to drug-loaded liposomes through a thioether connection between the free thiol of the antibody fragment and the MAL group. Alternately, the terminal MAL group on the liposomes' surface was conjugated to the thiol group. Using an ammonium sulfate gradient, doxorubicin was injected into immunoliposomes that were already produced (Park et al., 2002).

4.3.2. Targeting of Folate Receptors (FRs)

The overexpression of Folate Receptors, which can be detected on the exterior of diverse types of cancer cells, such as those found in the lungs and breasts, has been documented. An investigation carried out by Low et al. in 2008 exhibited that doxorubicin liposomes that were specifically directed towards Folate Receptors displayed a higher degree of cytotoxicity in comparison to conventional doxorubicin liposomes. In order to fabricate the liposomes that were aimed at Folate, Folate was linked to the liposomes through the utilization of a PEG spacer. This milestone was achieved through the incorporation of PEGylated lipid into the liposomes (Low et al., 2008).

4.3.3. Targeting of Transferrin Receptors (TfRs)

Transferrin (Tf) serves as a transport protein for Fe^{3+} ions in the serum, possessing an approximate molecular weight of 80 kilodaltons. The presence of TfRs (TfR1 and TfR2 or CD77) on the cell surface signifies their role as receptors for transferrin (Tf). The Tf-TfR complex is internalized through the process of endocytosis. In response to the increased iron demands displayed by cancerous cells, the expression of transferrin receptors is upregulated within the cancer context. The use of TfR-targeted doxorubicin liposomes against liver cancer has shown greater therapeutic effectiveness. Comparing docetaxel-containing TfR-targeted liposomes to docetaxel-containing non-targeted liposomes, a markedly greater degree of cytotoxicity was found (Heath et al., 2013; Li et al., 2009; Zhai et al., 2010).

4.4. Tumor Microenvironments Targeting

Another strategy involves the selective targeting of overexpressed receptors in the tumor microenvironment, tumor vasculature, and endothelial cells of tumor neovasculature, particularly within the blood vessels of the tumor. The development of novel blood vessels is of utmost importance in order to procure the requisite blood supply for the proliferation of the tumor. The disruption of the vasculature, denoting the organization of blood vessels within an organ, impedes the propagation of malignant cells. Various receptors are excessively expressed within the tumor microenvironment, and these can be selectively aimed at in order to efficiently transport anticancer agents to the desired site (Linton et al., 2016).

4.4.1. Targeting of Vascular Cell-Adhesion Molecules (VCAMs)

The components mentioned show a crucial part in the inflammatory process. In research conducted by Chiu and colleagues, the investigation centered on VCAM-1, that is overexpressed antigen, in both non-small cell lung cancer (NSCLC) cells and cancer vasculature. The investigation meticulously analyzed the binding capability of immunoliposomes targeting VCAM-1. As per Chiu et al.'s research in 2003, the binding affinity of antibody-conjugated liposomes was found to be eight times greater compared to unconjugated antibody liposomes. Another research study demonstrated the accumulation of VCAM-1 targeted liposomes within tumor blood vessels in animal models with tumor xenografts (Gosk et al., 2008).

4.4.2. Targeting of Integrins

These glycoproteins, which span the cell membrane, are over-expressed in tumor cells that form new blood vessels in the epithelial tissue. An amino acid sequence consisting of three residues, RGD (Arg-Gly-Asp), exhibits a robust attraction to integrins. Liposomes that underwent conjugation with RGD were formulated to precisely target integrins. These RGD-conjugated liposomes, encapsulating paclitaxel, exhibited a heightened drug concentration within tumor cells in comparison to non-targeted paclitaxel liposomes (Meng et al., 2011). Doxorubicin liposomes, when conjugated with RGD, exhibited an elevated cellular absorption of the therapeutic compound in the U87MG cellular lineage when compared to the uncomplicated doxorubicin liposomes (Chen et al., 2012).

4.4.3. Targeting of Matrix-Metalloproteases

Matrix-Metalloproteases (MMPs) represent a group of proteins, known as enzymes, that have the capability to degrade the extracellular matrix (ECM). Additionally, these enzymes are involved in the process of angiogenesis, which is the formation of new blood cells. Tumor tissues display the expression of various MMPs, with MT1-MMP being particularly prominent. In the context of HT 1080 cancer cells that exhibit over-expression of MT1-MMP, liposomes containing doxorubicin and anti-MT1-MMP Fab fragments demonstrated an increased cellular uptake compared to non-targeted liposomes (Hatakeyama et al., 2007).

4.4.4. Targeting of Cluster-of-Differentiation 44 (CD44)

Transmembrane adhesion molecule CD44 performs the role of a receptor protein. Its presence is heightened in a multitude of tumors including colon, breast, and others. Owing to its specific affinity for hyaluronic acid, a modality was devised to modify mesoporous silica nano particles by incorporating doxorubicin. The aim was to specifically target CD44. Notably, the cancer cells exhibited enhanced cytotoxicity when compared to non-targeted silica particles (Yu et al., 2013). In a distinct study, liposomes were synthesized incorporating anti-CD44 aptamer-1 by conjugating deprotected Apt1 to the terminal MAL group in preexisting liposomes. This conjugation process occurred through an overnight incubation at a temperature of 4°C. The resulting aptamer-1 liposomes exhibited a strong affinity for cancer cells that expressed the CD44 protein (Alshaer et al., 2015).

5. Applications of Liposome-Based Drugs for Different Types of Cancer Therapy

Chemotherapeutics represent the primary treatment modality employed for combating cancer. Nevertheless, their utility is significantly curtailed owing to the overt manifestation of toxic effects, inadequacy in targeting specific tissues, a narrow range of efficacy, and the heightened propensity for drug resistance. These factors concomitantly contribute to a substantial failure in the management of cancer. However, it has been shown that the development of nanoscale liposomal formulations makes it easier to deliver therapeutic medicines to tumor cells specifically, avoiding the non-specific toxicity brought on by the increased permeability and retention (EPR) effect (Olusanya et al., 2018).

Doxorubicin (DOX) is a bactericide produced from the anthracycline known as Streptomyces peucetius var. caesius and has strong anti-cancer activities. It is frequently used in the treatment of breast cancer and lymphoma as well as other solid and hematologic neoplasms. However, significant cardiotoxicity and cytotoxicity limit its clinical application. Cardiomyopathy and congestive heart failure are often irreversibly damaged as a result of the cumulative consequences of cardiac toxicity. The production of free radicals and lipid peroxidation are the causes of these consequences. DOX affects DNA through various mechanisms, primarily by inserting into the minor groove of the DNA double helix due to electrostatic interactions between sugar moieties and phosphate residues. It also stabilizes DNA-topoisomerase II, hindering DNA resealing and cell replication, and can induce apoptosis, often triggered by the DNA break repair process (Tahover et al., 2015; Mostafa et al., 2000).

Daunorubicin (DNR) is an additional antibiotic belonging to the anthracycline class. It is derived from the bacterial strain Streptomyces peucetius varcaesitue and possesses anticancer properties. While causing significant adverse effects like dose-dependent cardiotoxicity, alopecia, nausea, and vomiting, its mechanism of action mirrors that of DOX. The adverse effects are notably associated with DNR treatment. However, a viable substitute to address some of these unfavorable effects has been devised for liposomal daunorubicin. Liposomal DNR, commercially known as DaunoXome®, measures 50 nm in size and constitutes a 2:1 molar ratio mixture of DSPC and cholesterol. In 1996, the FDA approved this formulation and named it the single first-line therapy for advanced Kaposi's sarcoma linked with HIV. It has been created particularly to be absorbed by bloodstream monocytes (Gill et al., 1996). DaunoXome®'s small size and absence of an electrical charge help to reduce reticuloendothelial system (RES) absorption, which in turn extends the time that the medication is in the bloodstream. When this formulation was used in therapeutic settings, treated mice had higher plasma levels of doxorubicin (DNR) than mice that received the medication without restrictions. Similarly, the unrestrained drug's removal rate was found to be 44.9 mL/h, but the liposomal formulation showed a much slower elimination rate of 0.195 mL/h, pointing to DaunoXome®'s poor clearance rate. DaunoXome® was shown to have a favorable cardiac toxicity profile in a separate trial with a correlated nature, which later allowed for an increase in the anthracycline dosage without concomitant increases in cardiotoxicity (Creutzig et al., 2013).

Mitoxantrone (MXT) is an antineoplastic drug of the anthracenedione class, frequently employed in the management of various malignancies including lymphomas, breast cancer, prostate cancers, and leukemias (Kucherov et al., 2017). Animal studies have provided evidence indicating that free MXT may possess cardiotoxic capabilities. Nevertheless, the impact on cardiac tissue caused by this pharmaceutical agent is less severe compared to that of free DOX, owing to its utilization of an alternative mechanism (Alderton et al., 1992). Ongoing investigations are currently in Phase II for MXT liposomes, designed to target lymphoma and breast cancer. Wang et al. (2010) compared the pharmacokinetics, pharmacodynamics, and tissue distribution of MXT-containing liposomes with free MXT in animal research. The study revealed that MXT-loaded liposomes effectively prevented tumor progression, as evidenced by the results of the pharmacodynamic experiments. Moreover, the research on the antitumor effects demonstrated a significantly enhanced therapeutic efficacy of the drug when using liposomes with loaded MXT compared to free MXT. Furthermore, the pharmacokinetic assessments demonstrated that, at an equivalent dosage, MXT-loaded liposomes exhibited prolonged circulation compared to free MXT. Last but not least, the results of the tissue distribution investigations indicated that MXT loaded-liposomes aggregated in tumor regions as opposed to healthy tissues. The use of MXT loaded-liposomes has the potential to improve the therapeutic impact and therapeutic index of MXT, according to the overall findings of these research (Wang et al., 2010).

Paclitaxel (PCX) is an organic compound that is derived from the plant species Taxus brevifolia (Zhou et al., 2013). The anticancer effectiveness of

PCX is constrained as a result of its significant lipophilicity. Additionally, its water solubility is exceedingly low, measuring less than 0.01 mg/mL (Surapaneni et al., 2012). PCX makes up the pharmaceutical formulation of Taxol®, which is given as a non-aqueous base vehicle. Due to the inclusion of Cremophor EL in the preparation, patients have had significant hypersensitivity responses as well as precipitation when the medication is diluted with water (Singla et al., 2002). A comparison study was conducted to assess the efficacy and safety of PCX liposome (Lipusu®) vs free PCX in patients with stomach cancer. Throughout the study, patients were administered additional chemotherapeutic medications commonly used for individuals with advanced gastric cancer. While the hematological and neurological toxicities of both preparations were almost identical, individuals who received PCX in the form of liposomal tablets had a notable reduction in side effects such nausea, hypersensitivity responses, and vomiting (Xu et al., 2013).

Wang-Gillam and colleagues (2018) provide a comprehensive report on the outcomes obtained from a global, phase 3, randomized, open-label trial conducted at 76 sites across 14 nations. Individuals diagnosed with metastatic pancreatic ductal adenocarcinoma were the target population of this experiment. The NAPOLI-1 trial aimed to evaluate the impact of irinotecan, either as a standalone or in conjunction with folinic acid and fluorouracil within a nanoliposomal formulation. When nanoliposomal irinotecan was incorporated into the fluorouracil and folinic acid combination, patients with prior gemcitabine-based therapy experienced a notable improvement in survival rates, presenting a distinct therapeutic avenue (Wang-Gillam et al., 2016).

Silver nanoparticles possess antitumor properties due to their capability to constrain cell growth and persuade cell death. This study aims to create liposomes loaded with silver nanoparticles to effectively treat cancer. Thin-film hydration and sonication methods were used to create the liposomes. The dialysis bag method was used to analyze the drug release behavior of the silver nanoparticle-loaded liposomes, and mathematical models were used to confirm the results. The liposomes exhibited a high encapsulation efficiency. The nanoparticles within the liposomes were spherical and ranged in size from 80-97 nm. The release of the drug from the liposomes was observed to vary with pH, with the highest release occurring at pH 5.5, a characteristic pH of tumor cells. The drug release data aligned best with the Higuchi model, suggesting a diffusion-controlled release mechanism. These findings suggest that silver nanoparticle-loaded liposomes have potential as a targeted drug delivery system for various types of cancer (Jayachandran et al., 2023).

A key tactic in immunotherapy is to target the tumor's immunosuppressive microenvironment. But the importance of the immunological milieu in tumor lymph nodes is frequently overlooked. In this work, researchers introduce a nanoinducer called NIL-IM-Lip that, by energizing T and NK cells in tumor lymph nodes, successfully changes the immune milieu that has been suppressed there. When applied to tumors, the temperature-sensitive NIL-IM-Lip travels to the lymph nodes and releases IL-15. IR780 and 1-MT cause immunogenic cell death and inhibit regulatory T cells under photothermal stimulation. When paired with anti-PD-1, the use of NIL-IM-Lip improves the effectiveness of both T and NK cells, resulting in decreased tumor development in both hot and cold tumor models. This study highlights the significance of the immune milieu within tumor lymph nodes in the context of immunotherapy and supports the combination of immune checkpoint blockage with lymph node targeting in cancer immunotherapy (Fu et al., 2023).

Insufficient penetration of tumors, rapid elimination, systemic toxicity, and the development of resistance to drugs can impose limitations on the effectiveness of therapeutic cancer medications. The adoption of liposomal platforms holds promise in addressing the suboptimal therapeutic index arising from challenges like inefficient drug penetration, rapid drug elimination, and toxicity. Co-administered with docetaxel, liposomes demonstrate potential in mitigating drug resistance to pemetrexed. Scientists have developed a specialized liposomal formulation capable of delivering docetaxel and pemetrexed concurrently, thereby enhancing therapy efficacy and safety. The liposomes effectively co-encapsulated hydrophobic docetaxel and hydrophilic pemetrexed. Consequently, these liposomes significantly heightened cytotoxicity and induced immunogenic cell death in cellular experiments. The medication combination based on liposomes inhibited tumor development *in vivo* and promoted immunological responses. The drugs' encapsulation in liposomes reduced their systemic toxicity as compared to conventional free medication delivery. For the treatment of colon malignancies, a promising approach is the combination of anti-PD-L1 immunotherapy, docetaxel, pemetrexed, and liposome-mediated chemotherapy (Gu et al., 2023).

Triple-negative breast cancer represents a challenging and life-threatening subtype of breast cancer. Herbal medicine is being used more often to treat TNBC because it is more effective and has fewer side effects than chemotherapy. Crocin, a substance found in saffron, has been shown to cause cell death in TNBC cells. A new formulation of crocin using liposomes was created and tested. The liposomes were found to be effective at delivering crocin to the cancer cells and increasing their death. When combined with the chemotherapy drug DOX, the liposomes had an even greater effect. This study suggests that using liposomes to deliver crocin could be a promising treatment for TNBC (Chavoshi et al., 2023).

6. Emerging Trends and Opportunities for Improving Liposome-Based Cancer Therapy

- Active targeting strategies: The development of methods for deliberately directing liposomes toward a tumor has been the major focus of subsequent work on enhancing the therapeutic potential of liposomes (Khan et al., 2020; Wang et al., 2023).
- Improved pharmacokinetic properties: Long circulation times, better pharmacokinetic qualities of the medication contained, and passive tumor targeting are just a few benefits that liposomes provide (Fulton and Missaoui, 2023).
- Enhanced drug solubility: Liposomes are able to enhance drug solubility, which is particularly important for chemotherapeutic drug delivery (Kumbham et al., 2023).
- Liposomal formulations: Studies have shown that employing liposomes to deliver medication formulations has significant benefits over first-line therapies already in use (Raza et al., 2022).
- Cancer immunotherapy: With its great efficiency in several applications, including cancer immunotherapy, liposomes have emerged as a viable method to address the shortcomings in existing cancer treatments (Gu et al., 2020).

7. Current Challenges and Limitations of Liposome-Based Cancer Therapies

Liposome based cancer therapies are a promising approach to deliver drugs selectively to tumor cells, while minimizing the toxicity to normal tissues. However, there are several challenges and limitations that need to be overcome to improve the efficacy and safety of liposome-based cancer therapies. Some of these challenges and limitations are:

- Stability and circulation time: Liposomes are prone to degradation by enzymes, pH changes, and oxidation in the blood stream. They can also be cleared by the immune system or the reticuloendothelial system (RES), which reduces their circulation time and availability to reach the tumor site. To enhance the stability and circulation time of liposomes, various strategies have been developed, such as coating them with polymers, lipids, or proteins, modifying their size and charge, or incorporating stealth molecules (Yang et al., 2021; Kieler et al., 2018).
- Targeting and penetration: Liposomes can be targeted to specific tumor cells by adding ligands that identify receptors or antigens on the surface of tumor cells, such as antibodies, peptides, or aptamers. However, targeting alone is not sufficient to ensure the delivery of drugs to tumor cells. Liposomes also need to overcome the barriers posed by the tumor microenvironment, such as high interstitial pressure, dense extracellular matrix, heterogeneous vascularization, and hypoxia. These factors can limit the penetration and distribution of liposomes within the tumor tissue. To improve the targeting and penetration of liposomes, various strategies have been developed, such as using stimuli-responsive liposomes that can release their contents in response to pH, temperature, light, or enzymes, using ultrasound or magnetic fields to enhance the permeability of tumor vessels or cells, or using combination therapies with other agents that can modulate the tumor microenvironment (Yang et al., 2021; Kieler et al., 2018).
- Drug release and efficacy: Liposomes can encapsulate various types of drugs, such as small molecules, proteins, nucleic acids, or nanoparticles. However, the release of drugs from liposomes can be affected by several factors, such as the stability of the liposomal membrane, the interaction between the drug and the liposome components, the diffusion rate of the drug across the liposomal membrane, and the presence of competing molecules in the extracellular or intracellular environment. The release of drugs from liposomes can also be influenced by the biological barriers encountered by liposomes during their journey from the injection

site to the tumor site. These factors can affect the bioavailability and efficacy of drugs delivered by liposomes. To optimize the drug release and efficacy of liposomes, various strategies have been developed, such as using controlled-release mechanisms that can regulate the release of drugs according to specific triggers or conditions, using synergistic combinations of drugs that can enhance their therapeutic effects, or using personalized approaches that can tailor the drug dosage and delivery according to individual patient characteristics (Yang et al., 2021; Kieler et al., 2018).

Conclusion

In conclusion, liposomes present a promising avenue for revolutionizing cancer therapy through their unique composition and drug delivery capabilities. The encapsulation of both hydrophilic and hydrophobic drugs within nanosized vesicles offers significant advantages, including enhanced drug stability, reduced toxicity, and improved pharmacokinetics. The ability to actively target cancer cells and utilize the enhanced permeability and retention effect amplifies the efficacy of treatment while minimizing damage to healthy tissues. This chapter has comprehensively explored the current state and prospects of liposome-based cancer therapy. Strategies for effective targeting, applications across different cancer types, and emerging trends have been discussed. Despite the promising potential of liposomes, challenges such as translational barriers and limitations in current therapies persist. Ongoing research, development, and clinical trials, including FDA-approved liposome-based drugs, are addressing these challenges, showcasing the evolving landscape of cancer treatment. While acknowledging these obstacles, the immense promise of liposome-based cancer therapy remains clear. As we continue to refine and innovate this technology, considering patient needs and optimizing drug formulations, liposomes are poised to play a pivotal role in enhancing patient outcomes and contributing to the advancement of cancer treatment on a broader scale. The journey from laboratory findings to clinical application requires collaborative efforts and sustained research to unlock the full potential of liposomes in the fight against cancer.

References

Aanisah N, Wardhana YW, Chaerunisaa AY, Budiman A. Review on modification of glucomannan as an excipient in solid dosage forms. *Polymers*. 2022 Jun 23;14(13):2550.

Abdelhamid M, Koutsamanis I, Corzo C, Maisriemler M, Ocampo AB, Slama E, Alva C, Lochmann D, Reyer S, Freichel T, Salar-Behzadi S. Filament-based 3D-printing of placebo dosage forms using brittle lipid-based excipients. *Int J Pharm*. 2022 Aug 25;624:122013.

Akbari J, Saeedi M, Ahmadi F, Hashemi SM, Babaei A, Yaddollahi S, Rostamkalaei SS, Asare-Addo K, Nokhodchi A. Solid lipid nanoparticles and nanostructured lipid carriers: A review of the methods of manufacture and routes of administration. *Pharm Dev Technol*. 2022 May 28;27(5):525-44.

Alderton PM, Gross J, Green MD. Comparative study of doxorubicin, mitoxantrone, and epirubicin in combination with ICRF-187 (ADR-529) in a chronic cardiotoxicity animal model. *Cancer Res*. 1992;52:194–201.

Alshaer W, Vergnaud J, Ismail S, Fattal E. Functionalizing Liposomes with anti-CD44 Aptamer for Selective Targeting of Cancer Cells. *Bioconjug Chem*. 2015;26:1307–1313.

Amin M, Seynhaeve AL, Sharifi M, Falahati M, Ten Hagen TL. Liposomal Drug Delivery Systems for Cancer Therapy: The Rotterdam Experience. *Pharmaceutics*. 2022 Oct 11;14(10):2165.

Arévalo-Pérez R, Maderuelo C, Lanao JM. Recent advances in colon drug delivery systems. *J Control Release*. 2020 Nov 10;327:703-24.

Aziz A, Rehman U, Sheikh A, Abourehab MA, Kesharwani P. Lipid-based nanocarrier mediated CRISPR/Cas9 delivery for cancer therapy. *J Biomater Sci Polym* Ed. 2022 Sep 3:1-21.

Barenholz Y. Doxil®—The first FDA-approved nano-drug: Lessons learned. *J Control Release*. 2012;160:117–134.

Benhabbour SR, Luft JC, Kim D, Jain A, Wadhwa S, Parrott MC, Liu R, DeSimone JM, Mumper RJ. *In vitro* and *in vivo* assessment of targeting lipid-based nanoparticles to the epidermal growth factor-receptor (EGFR) using a novel Heptameric ZEGFR domain. *J Control Release*. 2012;158:63–71.

Bigham A, Rahimkhoei V, Abasian P, Delfi M, Naderi J, Ghomi M, Moghaddam FD, Waqar T, Ertas YN, Sharifi S, Rabiee N. Advances in tannic acid-incorporated biomaterials: Infection treatment, regenerative medicine, cancer therapy, and biosensing. *Chem Eng J*. 2022 Mar 15;432:134146.

Boztepe T, Scioli-Montoto S, Gambaro RC, Ruiz ME, Cabrera S, Alemán J, Islan GA, Castro GR, León IE. Design, Synthesis, Characterization, and Evaluation of the Anti-HT-29 Colorectal Cell Line Activity of Novel 8-Oxyquinolinate-Platinum (II)-Loaded Nanostructured Lipid Carriers Targeted with Riboflavin. *Pharmaceutics*. 2023 Mar 22;15(3):1021.

Chavda VP, Vihol D, Mehta B, Shah D, Patel M, Vora LK, Pereira-Silva M, Paiva-Santos AC. Phytochemical-loaded liposomes for anticancer therapy: An updated review. *Nanomedicine*. 2022 Apr;17(8):547-68.

Chavoshi H, Taheri M, Wan ML, Sabzichi M. Crocin-loaded liposomes sensitize MDA-MB 231 breast cancer cells to doxorubicin by inducing apoptosis. *Process Biochemistry*. 2023 Jul 1;130:272-80.

Chen M, Wang S, Qi Z, Meng X, Hu M, Liu X, Song Y, Deng Y. Deuterated colchicine liposomes based on oligomeric hyaluronic acid modification enhance anti-tumor effect and reduce systemic toxicity. *Int J Pharm.* 2023 Feb 5;632:122578.

Chen Z, Deng J, Zhao Y, Tao T. Cyclic RGD peptide-modified liposomal drug delivery system: Enhanced cellular uptake *in vitro* and improved pharmacokinetics in rats. *Int J Nanomed*. 2012;7:3803–3811.

Chiu GNC, Bally MB, Mayer LD. Targeting of antibody conjugated, phosphatidylserine-containing liposomes to vascular cell adhesion molecule 1 for controlled thrombogenesis. *Biochim Biophys Acta Biomembr*. 2003;1613:115–121.

Cho HJ. Recent progresses in the development of hyaluronic acid-based nanosystems for tumor-targeted drug delivery and cancer imaging. *J Pharm Investig*. 2020 Mar;50:115-29.

Creutzig U, Zimmermann M, Bourquin JP, Dworzak MN, Fleischhack G, Graf N, Klingebiel T, Kremens B, Lehrnbecher T, von Neuhoff C, Ritter J. Randomized trial comparing liposomal daunorubicin with idarubicin as induction for pediatric acute myeloid leukemia: results from Study AML-BFM 2004. Blood, *The Journal of the American Society of Hematology*. 2013 Jul 4;122(1):37-43.

Crivelli B, Perteghella S, Bari E, Sorrenti M, Tripodo G, Chlapanidas T, Torre ML. Silk nanoparticles: From inert supports to bioactive natural carriers for drug delivery. *Soft Matter*. 2018;14(4):546-57.

Dhaval M, Vaghela P, Patel K, Sojitra K, Patel M, Patel S, Dudhat K, Shah S, Manek R, Parmar R. Lipid-based emulsion drug delivery systems—a comprehensive review. *Drug Deliv Transl Res*. 2022 Jul 1:1-24.

Fath IS, Oyelere AK. Liposomal drug delivery systems for targeted cancer therapy: Is active targeting the best choice? *Future Med Chem.* 2016;8:2091–2112.

Feng L, Mumper RJ. A critical review of lipid-based nanoparticles for taxane delivery. *Cancer Lett*. 2013;334:157–175.

Fu S, Chang L, Liu S, Gao T, Sang X, Zhang Z, Mu W, Liu X, Liang S, Yang H, Yang H. Temperature sensitive liposome based cancer nanomedicine enables tumour lymph node immune microenvironment remodelling. *Nature Communications*. 2023 Apr 19;14(1):2248.

Fulton MD, Najahi-Missaoui W. Liposomes in Cancer Therapy: How Did We Start and Where Are We Now. *Int J Mol Sci*. 2023 Apr 1;24(7):6615.

Gao Y, Li J, Zhao L, Hong Y, Shen L, Wang Y, Lin X. Distribution pattern and surface nature-mediated differential effects of hydrophilic and hydrophobic nano-silica on key direct compaction properties of Citri Reticulatae Pericarpium powder by co-processing. *Powder Technol.* 2022 May 1;404:117442.

Gao Y, Zeng Y, Liu X, Tang D. Liposome-mediated in situ formation of type-I heterojunction for amplified photoelectrochemical immunoassay. *Anal Chem.* 2022 Mar 9;94(11):4859-65.

Garcia MA, Garcia CF, Faraco AA. Pharmaceutical and biomedical applications of native and modified starch: A review. *Starch-Stärke*. 2020 Jul;72(7-8):1900270.

Gill PS, Wernz J, Scadden DT, Cohen P, Mukwaya GM, von Roenn JH, Jacobs M, Kempin S, Silverberg I, Gonzales G, Rarick MU. Randomized phase III trial of liposomal daunorubicin versus doxorubicin, bleomycin, and vincristine in AIDS-related Kaposi's sarcoma. *Journal of Clinical Oncology*. 1996 Aug;14(8):2353-64.

Gosk S, Moos T, Gottstein C, Bendas G. VCAM-1 directed immunoliposomes selectively target tumor vasculature *in vivo*. *Biochim Biophys Acta Biomembr*. 2008;1778:854–863.

Gu Z, Da Silva CG, Van der Maaden K, Ossendorp F, Cruz LJ. Liposome-Based Drug Delivery Systems in Cancer Immunotherapy. *Pharmaceutics*. 2020;12:1054.

Gu Z, Da Silva CG, Hao Y, Schomann T, Camps MG, van der Maaden K, Liu Q, Ossendorp F, Cruz LJ. Effective combination of liposome-targeted chemotherapy and PD-L1 blockade of murine colon cancer. *Journal of Controlled Release*. 2023 Jan 1;353:490-506.

Gugleva V, Andonova V. Recent Progress of Solid Lipid Nanoparticles and Nanostructured Lipid Carriers as Ocular Drug Delivery Platforms. *Pharmaceuticals*. 2023 Mar 22;16(3):474.

Hatakeyama H, Akita H, Ishida E, Hashimoto K, Kobayashi H, Aoki T, Yasuda J, Obata K, Kikuchi H, Ishida T, Kiwada H. Tumor targeting of doxorubicin by anti-MT1-MMP antibody-modified PEG liposomes. *International journal of pharmaceutics*. 2007 Sep 5;342(1-2):194-200.

Heath JL, Weiss JM, Lavau CP, Wechsler DS. Iron deprivation in cancer-potential therapeutic implications. *Nutrients*. 2013;5:2836–2859.

Jayachandran P, Ilango S, Suseela V, Nirmaladevi R, Shaik MR, Khan M, Khan M, Shaik B. Green Synthesized Silver Nanoparticle-Loaded Liposome-Based Nanoarchitectonics for Cancer Management: *In Vitro* Drug Release Analysis. *Biomedicines*. 2023;11:217.

Khan MI, Hossain MI, Hossain MK, Rubel MH, Hossain KM, Mahfuz AM, Anik MI. Recent progress in nanostructured smart drug delivery systems for cancer therapy: a review. *ACS Appl Bio Mater*. 2022 Feb 28;5(3):971-1012.

Khan AA, Allemailem KS, Almatroodi SA, Almatroudi A, Rahmani AH. Recent strategies towards the surface modification of liposomes: an innovative approach for different clinical applications. *3 Biotech*. 2020 Apr;10:1-5.

Kieler M, Unseld M, Bianconi D, Prager G. Challenges and Perspectives for Immunotherapy in Adenocarcinoma of the Pancreas: The Cancer Immunity Cycle. *Pancreas*. 2018 Feb;47(2):142-157.

Kucherov FA, Egorova KS, Posvyatenko AV, Eremin DB, Ananikov VP. Investigation of cytotoxic activity of mitoxantrone at the individual cell level by using ionic-liquid-tag-enhanced mass spectrometry. *Anal Chem.* 2017;89:13374–13381.

Kumbham S, Ajjarapu S, Ghosh B, Biswas S. Current trends in the development of liposomes for chemotherapeutic drug delivery. *Journal of Drug Delivery Science and Technology*. 2023 Aug 16:104854.

Kurakula M, Gorityala S, Moharir K. Recent trends in design and evaluation of chitosan-based colon targeted drug delivery systems: Update 2020. *J Drug Deliv Sci Technol*. 2021 Aug 1;64:102579.

Lazić I, Kučević S, Ćirin-Varađan S, Aleksić I, Đuriš J. Formulation of ibuprofen-modified release hydrophilic and lipid matrix tablets using co-processed excipients. *Arhiv za farmaciju*. 2022;72(4 suplement):S3400-401.

Li X, Ding L, Xu Y, Wang Y, Ping Q. Targeted delivery of doxorubicin using stealth liposomes modified with transferrin. *Int J Pharm*. 2009;373:116–123.

Li Z, Yin Z, Li B, He J, Liu Y, Zhang N, Li X, Cai Q, Meng W. Docosahexaenoic Acid-Loaded Nanostructured Lipid Carriers for the Treatment of Peri-Implantitis in Rats. *Int J Mol Sci*. 2023 Jan;24(3):1872.

Lim SB, Banerjee A, Önyüksel H. Improvement of drug safety by the use of lipid-based nanocarriers. *J Control Release*. 2012;163:34–45.

Linton SS, Sherwood SG, Drews KC, Kester M. Targeting cancer cells in the tumor microenvironment: Opportunities and challenges in combinatorial nanomedicine. *Wiley Interdiscip Rev Nanomed Nanobiotechnol*. 2016;8:208–222.

Liu P, Chen G, Zhang J. A review of liposomes as a drug delivery system: current status of approved products, regulatory environments, and future perspectives. *Molecules*. 2022 Feb 17;27(4):1372.

Low PS, Henne WA, Doorneweerd DD. Discovery and development of folic-acid-based receptor targeting for imaging and therapy of cancer and inflammatory diseases. *Acc Chem Res*. 2008;41:120–129.

Luiz MT, Dutra JA, Tofani LB, de Araújo JT, Di Filippo LD, Marchetti JM, Chorilli M. Targeted liposomes: a nonviral gene delivery system for cancer therapy. *Pharmaceutics*. 2022 Apr 8;14(4):821.

Maeda H, Nakamura H, Fang J. The EPR effect for macromolecular drug delivery to solid tumors: Improvement of tumor uptake, lowering of systemic toxicity, and distinct tumor imaging *in vivo*. *Adv Drug Deliv Rev*. 2013;65:71–79.

Maheshwari S. AGEs RAGE Pathways: Alzheimer's Disease. *Drug Res*. 2023 Mar 20.

Mamot C, Drummond DC, Noble CO, Kallab V, Guo Z, Hong K, Kirpotin DB, Park JW. Epidermal growth factor receptor-targeted immunoliposomes significantly enhance the efficacy of multiple anticancer drugs *in vivo*. *Cancer Res*. 2005;65:11631–11638.

Meng S, Su B, Li W, Ding Y, Tang L, Zhou W, Song Y, Caicun Z. Integrin-targeted paclitaxel nanoliposomes for tumor therapy. *Med Oncol*. 2011;28:1180–1187.

Mohammadi M, Hamishehkar H, McClements DJ, Shahvalizadeh R, Barri A. Encapsulation of Spirulina protein hydrolysates in liposomes: Impact on antioxidant activity and gastrointestinal behavior. *Food Chem*. 2023 Jan 30;400:133973.

Moosavian SA, Kesharwani P, Singh V, Sahebkar A. Aptamer-functionalized liposomes for targeted cancer therapy. *Aptamers Engineered Nanocarriers for Cancer Therapy*. 2023 Jan 1:141-72.

Mostafa MG, Mima T, Ohnishi ST, Mori K. S-allylcysteine ameliorates doxorubicin toxicity in the heart and liver in mice. *Planta Med*. 2000;66:148–151.

Mouhid L, Corzo-Martínez M, Torres C, Vázquez L, Reglero G, Fornari T, Ramírez de Molina A. Improving *in vivo* efficacy of bioactive molecules: An overview of potentially antitumor phytochemicals and currently available lipid-based delivery systems. *J Oncol*. 2017 May 7;2017.

Nakmode D, Bhavana V, Thakor P, Madan J, Singh PK, Singh SB, Rosenholm JM, Bansal KK, Mehra NK. Fundamental aspects of lipid-based excipients in lipid-based product development. *Pharmaceutics*. 2022 Apr 11;14(4):831.

Nguyen TT, Maeng HJ. Pharmacokinetics and pharmacodynamics of intranasal solid lipid nanoparticles and nanostructured lipid carriers for nose-to-brain delivery. *Pharmaceutics*. 2022 Mar 5;14(3):572.

Olusanya TOB, Haj Ahmad RR, Ibegbu DM, Smith JR, Elkordy AA. Liposomal Drug Delivery Systems and Anticancer Drugs. *Molecules*. 2018 Apr 14;23(4):907.

Park JW, Hong K, Kirpotin DB, Colbern G, Shalaby R, Baselga J, Shao Y, Nielsen UB, Marks JD, Moore D, Papahadjopoulos D. Anti-HER2 immunoliposomes: enhanced efficacy attributable to targeted delivery. *Clinical Cancer Research*. 2002 Apr 1;8(4):1172-81.

Radmoghaddam ZA, Honarmand S, Dastjerdi M, Akbari S, Akbari A. Lipid-based nanoformulations for TKIs delivery in cancer therapy. *NanoScience Technology*. 2022:11-27.

Rahman M, Almalki WH, Alrobaian M, Iqbal J, Alghamdi S, Alharbi KS, Alruwaili NK, Hafeez A, Shaharyar A, Singh T, Waris M. Nanocarriers-loaded with natural actives as newer therapeutic interventions for treatment of hepatocellular carcinoma. *Expert Opin Drug Deliv*. 2021 Apr 3;18(4):489-513.

Raza F, Evans L, Motallebi M, Zafar H, Pereira-Silva M, Saleem K, Peixoto D, Rahdar A, Sharifi E, Veiga F, Hoskins C. Liposome-based diagnostic and therapeutic applications for pancreatic cancer. *Acta Biomaterialia*. 2022 Dec 12.

Riaz MK, Riaz MA, Zhang X, Lin C, Wong KH, Chen X, Zhang G, Lu A, Yang Z. Surface Functionalization and Targeting Strategies of Liposomes in Solid Tumor Therapy: A Review. *Int J Mol Sci*. 2018 Jan 9;19(1):195.

Singh A, Ansari VA, Ahsan F, Akhtar J, Khushwaha P, Maheshwari S. Viridescent concoction of genstein tendentious silver nanoparticles for breast cancer. *Res J Pharm Technol*. 2021;14(5):2867-72.

Singh A, Ansari VA, Haider F, Akhtar J, Ahsan F. A Review on Topical preparation of Herbal Drugs used in Liposomal delivery against Ageing. *Res J Pharmacol Pharmacodyn*. 2020;12(1):5-11.

Singh A, Ansari VA, Mahmood T, Ahsan F, Wasim R, Shariq M, Parveen S, Maheshwari S. Receptor for Advanced Glycation End Products: Dementia and Cognitive Impairment. *Drug Res*. 2023 Mar 8.

Singh A, Ansari VA, Mahmood T, Ahsan F, Wasim R. Neurodegeneration: Microglia: Nf-Kappab Signaling Pathways. *Drug Res*. 2022 Sep 2.

Singla AK, Garg A, Aggarwal D. Paclitaxel and its formulations. *Int J Pharm*. 2002;235:179–192.

Surapaneni MS, Das SK, Das NG. Designing paclitaxel drug delivery systems aimed at improved patient outcomes: Current status and challenges. *ISRN Pharmacol*. 2012;2012:623139.

Tahover E, Patil YP, Gabizon AA. Emerging delivery systems to reduce doxorubicin cardiotoxicity and improve therapeutic index: focus on liposomes. *Anticancer Drugs*. 2015;26:241–258.

Torchilin VP. Passive and active drug targeting: Drug delivery to tumors as an example. In: Schäfer-Korting M., editor. *Drug Delivery*. Springer; Berlin/Heidelberg, Germany: 2010. pp. 3–53.

Tranová T, Macho O, Loskot J, Mužíková J. Study of rheological and tableting properties of lubricated mixtures of co-processed dry binders for orally disintegrating tablets. *Eur J Pharm Sci*. 2022 Jan 1;168:106035.

Van NH, Vy NT, Van Toi V, Dao AH, Lee BJ. Nanostructured lipid carriers and their potential applications for versatile drug delivery via oral administration. *OpenNano*. 2022 Aug 6:100064.

Wang CX, Li CL, Zhao X, Yang HY, Wei N, Li YH, Zhang L, Zhang L. Pharmacodynamics, pharmacokinetics and tissue distribution of liposomal mitoxantrone hydrochloride. *Yao Xue Xue Bao.* 2010;45:1565–1569.

Wang S, Chen Y, Guo J, Huang Q. Liposomes for Tumor Targeted Therapy: A Review. *Int J Mol Sci*. 2023, 24, 2643.

Wang-Gillam A, Li CP, Bodoky G, Dean A, Shan YS, Jameson G, Macarulla T, Lee KH, Cunningham D, Blanc JF, Hubner RA. Nanoliposomal irinotecan with fluorouracil and folinic acid in metastatic pancreatic cancer after previous gemcitabine-based therapy (NAPOLI-1): a global, randomised, open-label, phase 3 trial. *The Lancet*. 2016 Feb 6;387(10018):545-57.

Wasim R, Mahmood T, Siddiqui MH, Ahsan F, Shamim A, Singh A, Shariq M, Parveen S. Aftermath of AGE-RAGE Cascade in the pathophysiology of cardiovascular ailments. *Life Sci*. 2022 Aug 5:120860.

Xu X, Wang L, Xu HQ, Huang XE, Qian YD, Xiang J. Clinical comparison between paclitaxel liposome (Lipusu®) and paclitaxel for treatment of patients with metastatic gastric cancer. *Asian Pac J Cancer Prev.* 2013;14:2591–2594.

Yang Y, Zuo S, Li L, Kuang X, Li J, Sun B, Wang S, He Z, Sun J. Iron-doxorubicin prodrug loaded liposome nanogenerator programs multimodal ferroptosis for efficient cancer therapy. *Asian J Pharm Sci.* 2021 Nov;16(6):784-793.

Yu M, Jambhrunkar S, Thorn P, Chen J, Gu W, Yu C. Hyaluronic acid modified mesoporous silica nanoparticles for targeted drug delivery to CD44-overexpressing cancer cells. *Nanoscale*. 2013;5:178–183.

Zhai G, Wu J, Yu B, Guo C, Yang X, Lee RJ. A transferrin receptor-targeted liposomal formulation for docetaxel. *J Nanosci Nanotechnol.* 2010;10:5129–5136.

Zhou J, Zhao WY, Ma X, Ju RJ, Li XY, Li N, Sun MG, Shi JF, Zhang CX, Lu WL. The anticancer efficacy of paclitaxel liposomes modified with mitochondrial targeting conjugate in resistant lung cancer. *Biomaterials*. 2013;34:3626–3638.

Zhu YS, Tang K, Lv J. Peptide–drug conjugate-based novel molecular drug delivery system in cancer. *Trends Pharmacol Sci*. 2021 Oct 1;42(10):857-69.

Zupančič O, Spoerk M, Paudel A. Lipid-based solubilization technology via hot melt extrusion: Promises and challenges. *Expert Opinion on Drug Delivery*. 2022 Sep 2;19(9):1013-32.

Zylberberg C, Matosevic S. Pharmaceutical liposomal drug delivery: A review of new delivery systems and a look at the regulatory landscape. *Drug Deliv*. 2016;23:3319–3329.

Chapter 4

Liposomes in Breast Cancer, Cervical Cancer and Ovarian Cancer Therapy: Recent Advancements and Future Perspectives

Nazneen Sultana[1]
and Seema Devi[2,*]
[1]Pharmaceutics and Pharmacokinetics Division, CSIR - Central Drug Research Institute, Lucknow, Uttar Pradesh, India
[2]Department of Radiation Oncology, Indira Gandhi Institute of Medical Sciences, Patna, Bihar, India

Abstract

Breast cancer has been known to be the most prevalent cancer type in females worldwide, enumerating 25.8% of all female cancer cases, while cervical cancer and ovarian cancer contribute 6.9% and 3.6% of all female cancer cases, respectively. Although remarkable breakthroughs have been achieved in the prognosis, aetiology, and treatment of breast cancer, ovarian cancer, and cervical cancer, there remains an eminent cause of female mortality around the world. Additionally, the lack of specificity in traditional chemotherapies results in systemic toxicity, eventually leading to the inheritance of multidrug resistance in cancerous tissue. Also, liposomes have been recognised to revolutionise cancer treatment through their extensive clinical administration. Liposomes have significant potential to overcome the limitations of traditional chemotherapies because of their specific characteristics, such as their small size, biodegradability, biocompatibility, immunogenicity, hydrophobic/hydrophilic nature and

* Corresponding Author's Email: doctorseema71@gmail.com.

In: Liposomes
Editors: Usama Ahmad and Anas Islam
ISBN: 979-8-89113-636-6

low toxicity. Liposomes enhance the stability and bioavailability of therapeutic molecules and reduce side effects by targeting therapeutics to the cancer site. Liposomes were the first nano-based drug carriers approved for clinical practice. Few liposome preparations loaded with anti-cancer candidates to combat breast cancer, ovarian cancer, and cervical cancer are already present on the market whereas enormous liposome preparations are under research and clinical trials. Therefore, the present chapter will emphasise existing challenges related to breast cancer, cervical cancer, and ovarian cancer treatment and discuss the most recent strategies used to design liposome-based preparations to target breast cancer, cervical cancer, and ovarian cancer including the preparation undergoing clinical trials.

Keywords: cancer, chemotherapy, liposomes, active targeting, passive targeting

1. Introduction

In today's world, nanotechnology has become one of the smartest tools to cure human cancers and has gained the attention of cancer nanomedicine. Researchers have developed novel nanocarriers with the knowledge of molecular alteration based on pathophysiology alongside the capacity of biomolecules to express on the cancer tissue (Allred et al., 2001). However, effective novel nanocarriers should only bind to the targeted cancer tissue/cells with the least exposure to noncancerous normal healthy tissue/cells. Recent advances in comprehending the progression of breast, ovarian, and cervical cancer through molecular mechanisms have increased drastically. These findings are potentially useful in the blooming of target-based nanocarriers for the treatment of ovarian, breast, and cervical cancer.

1.1. Liposome: A Brief Account

A British hematologist first described liposomes in 1961 which was published a few years later in 1964 at the Babraham Institute, Cambridge. Dr. Alex and R. W. Horne modified the new electron microscope to incorporate a negative stain for dry phospholipids. It is obtained from the Greek words 'Lipo' meaning fat and "somes" meaning body; (body containing fat) (Bangham et al., 1964). Liposomes are single- or multi lipid bilayer structures surrounding an aqueous core, which can spontaneously form from the dispersion of amphiphilic lipids in water. Based on the size of the vesicles and the layer of phospholipid, liposomes can be multilamellar vesicles (multiple layers) (up to 5 μm), small unilamellar vesicles (single layer) (20-100 μm) and large uni-lamellar vesicle (single layer) (100-200 μm) (Nsairat et al., 2022). The similarity of liposomal structure with that of cellular membranes gives it an immense advantage in terms of biodegradation and biocompatibility. Additionally, liposomes can carry lipophilic, hydrophilic, and amphiphilic compounds either in their lipid bilayer or/and within the aqueous core (Nsairat et al., 2022; Elkhoury et al., 2020; Arab-Tehrany et al., 2020). Increases the stability of the loaded drug and acts as a protective agent by shielding drugs from heat, pH, and fluctuation of ions and avoiding enzymatic, environmental, and chemical changes. Various routes of administration have been developed for liposomal formulation to improve its remedial efficacy and patient compliance, such as ophthalmic, nasal, parenteral, transdermal, oral and pulmonary routes (Taha et al., 2014; Han Y, et al., 2019; Mirtaleb et al., 2021; Ferguson et al., 2023; Kurano et al., 2022). Additionally, liposomes have also been used in cosmetics (Kapoor et al., 2018) and the food industry (Ajeeshkumar et al., 2021) on a wide scale. Furthermore, advancement in liposomal engineering has revealed that liposome design may have different types of surface ligands that can attach to unhealthy tissue. And make it a more specific system for carrying therapeutic agents, reducing their toxicity and enabling their release in controlled manner to favour their clinical application (AlSawaftah et al., 2021; Antoniou et al., 2021; Bilal et al., 2021).

1.2. Liposomes in the Treatment of Breast Cancer

Globally, breast carcinoma is the most prevalent type of cancer with an incidence of over 2 million per year (Sung et al., 2020). New Zealand and Australia show the highest incidence worldwide, with more than 200,000 new cases per year (Sung et al., 2020; Australian Institute of Health and Welfare, 2021). Breast cancer is the leading cause of causality among total cancer-related mortality, which has resulted in about 685,000 causalities worldwide at the end of the year 2020 (Sung et al., 2020). The most preferred choice of treatment for breast cancer is chemotherapy; however, it has several restrictions such as a specificity deficit that often lead to drug toxicity and shows short- to long-term side effects (Mokhtari-Hessari et al.,

2018). Frequent chemotherapies in the patient where a particular therapy regime is essentially recommended is also a matter of cancer, since it not only causes physical and physiological implications but also imposes financial burden on patients (Wadhwani et al., 2020). Furthermore, chemotherapy is also related to multidrug resistance (MDR), which is the major cause of refractory, unresponsive, and recurrent cancers in most cancer patients. MDR causes overexpression of specific efflux transporters (ATP-binding cassette) which expulse API from the cancer cells before their onset of action. This process not only causes resistance in cancer for administered drugs but also for therapeutic moieties having similar mechanisms of action and molecular structure (Li Y-J et al., 2017; Press et al., 2017). Another challenge associated with chemotherapy is the metastatic progression of cancerous tissue to other organs of the body, especially the brain, bone, and lungs. Approximately 30-40% of breast cancer patients also suffer from metastatic tumors in the future and are more likely to succumb to them than primary cancer (Cardoso et al., 2015). To overcome these challenges, nanobased colloidal systems carry therapeutic components such as liposomes, dendrimers, and lipid nanocapsules have become the epicentre of the research recently. These colloidal carriers can carry drug molecules through various mechanisms. They can dissolve, encapsulate, covalently bond, embed, and adsorb along with the incorporation of functionalised antibodies, aptamers, peptides, proteins, ligands, and antigens that can target the cancer cell without/with minimum damage to healthy cells. They not only improve the pharmacokinetic characteristics but also enhance the safety of the therapeutic moiety, minimise its side effects, and enhance drug solubility (Di paolo et al., 2004; Miele et al., 2009; Eloy et al., 2014; Ansari et al., 2017; Kanwal et al., 2018). A diagrammatic representation of the different entry and its dissemblance by breast cancer cells is sketched and labeled in Figure 1.

2. Receptor-Mediated Targeting of Breast Cancer

In recent years, the preparation of a carrier system that can target therapeutics to malignant cells to assist in the therapy of breast cancer cases has attracted a lot of research attention in the area of novel drug development. Liposomes can actively and passively target breast tissue/cells. They can be used to successfully and efficiently target cancer tissue by enhanced permeability and retention effect (Muggia et al., 1999). The accumulation of target-based liposomal preparation is seen as much higher when compared to nontargeted liposomal preparation of breast cancer cells. An illustration of the penetration and processing of several liposomes in breast cancer cells is demonstrated in Figure 1. Molecular approaches are used to directly target tumor cells by active targeting through the interaction with cancer explicit markers (Lasic et al., 1998; Lasic et al., 1996). Actively targeting liposome systems are commonly formed by conjugating liposomes with antibodies or ligands such as folates, monoclonal antibodies, transferrin, etc. (Xu et al., 2013).

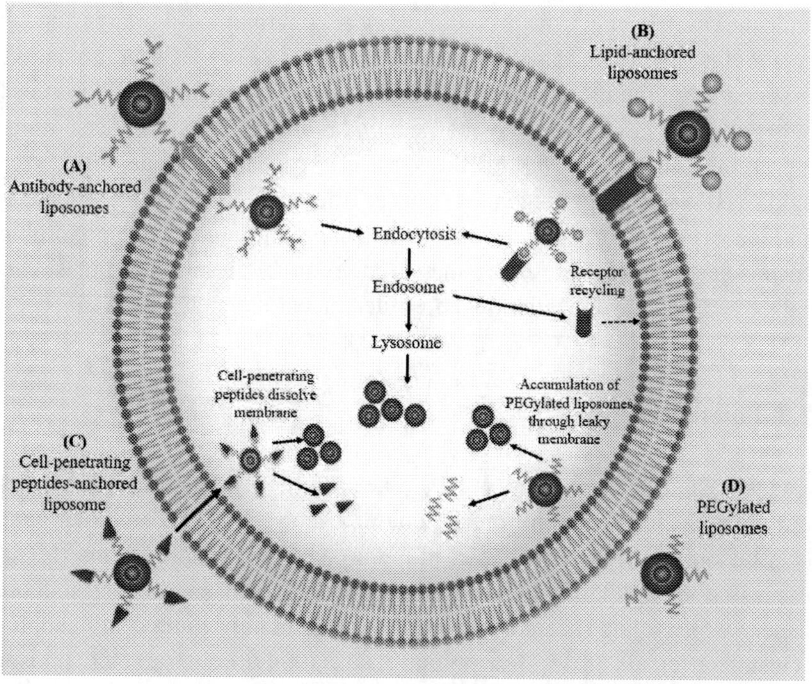

Figure 1. Bio-events in breast cancer cells for uptake and processing of different liposomes.

2.1. HER2 Receptor

Human epidermal growth factor receptor 2 (HER2) is underexpressed in normal tissue whereas abundantly expressed in most tissue of breast, ovarian, and gastric cancer, etc. The coupling agent targeting method is used for target drug delivery in cancer (Stroock et al., 2002). HER2 can be readily accessible receptors found on the outer surface of the tissue. It is highly expressed in 20% to 30% of breast cancer tissue and can be easily used as the site of immunotargeting of drug-delivering systems (Goyal et al., 2005; Park et al., 2001). HER2-based immunotargeting liposomes are prepared by sterically stable liposomal preparation and link it with scFv or Fab' of anti-HER2 antibodies. These then conjugate with the HER2 receptors present in breast cancer tissue and further induce intracellular accumulation of anticancer drugs (Hung et al., 1995). Trastuzumab and anti-HER2 antibody combined with paclitaxel-loaded liposomes preferentially increase the anticancer efficiency of drugs in cancer treatment by increasing drug accumulation in cancer cells (Zhigaltsev et al., 2012). The combination of cyst chain fragments of anti-HER2 antibody with doxorubicin liposomes has better tumour control due to the accumulation of drugs in breast cancers. Henceforth, it does not fall off the surface of cancerous tissue, and it can be employed as a target agent in cancer therapy. Vinu et al. showed 19 antibodies (Bernier-Latmani et al., 2017) combined with doxorubicin-encapsulated liposomes, which increased doxorubicin levels in breast cancer cells by rapid internalization in a dose-dependent manner.

2.2. Folate Receptors

Endocytosis folate receptors (FR) are not commonly found in normal tissue, but are prevalent in brain, breast, colon, kidney, lung, and ovarian cancers (Gustafson et al., 2015; Zhao et al., 2020). Folate receptor alpha (FR-α) is expressed in epithelial tumours like epithelial ovarian cancer, and breast cancer while folate receptor n-beta (FR-nβ) is mainly limited to nanophages (Gustafson et al., 2015; Harashima et al., 1994). Folic-targeted liposomal formulations can be used to carry chemotherapeutic drugs, genes, antigens oligonucleotides, and radio nucleotides to tumour cells with palate reception overexpression (Huckaby et al., 2018). Wu et al. used FR-targeted paclitaxel and found that guanidine cytotoxicity and layer terminal half-life showed as compound 70 non-targeted paclitaxel. Folic acid-conjugated liposomes have shown high anticancer activity through their increased circulating life and drug solubility (Badran et al., 2022). Rait and his co-workers observed in his study that the FR-targeted cationic liposomal preparation performed better in the *in vitro* study compared to non-targeting preparation in delivering anti-HER2 oligonucleotide to breast cancer cells (Rait et al., 2002). Long-term systemic circulation with significant accumulation of oligonucleotide was observed in breast cancer cells.

2.3. Transferrin Receptors

Transferrin is a glycoprotein that binds non-heme iron and is mainly required to transfer ferric iron to the plasma membrane through transferrin receptors. Since malignant cells require a substantial amount of iron, transferrin receptors are overexpressed superficially on the cancer cell. Transferrin receptors are expressed tenfold higher on tumor cell surfaces like breast cancer cells, leukemia cells, etc. If these receptors combine with stealth liposomes, it can increase the accumulation into breast cancer cells and inhibit cancer growth (Tonglairoum et al., 2016). The transferrin-conjugated liposomal preparation of doxorubicin and verapamil had shown better drug uptake by breast cancer cells and improved cytotoxicity (Wu et al., 2007).

2.4. Oestrogen Receptors

Oestrogen receptors are specialised proteins present on the outer surface of breast cancer tissue that bind to estrogen, a steroid hormone that facilitates the proliferation and progression of breast cancer tissue. Binding of oestrogen to ERs results in a cascade of signaling events that govern the expression of genes that are essential for cell growth, metabolism, and survival. Recent studies have shown that there are two main subtypes of ER that are associated with breast cancer. ER-α and ER-β. Wherein, ER-α is expressed primarily in the luminal breast cancer tissue, which almost accounts for 70% of total breast cancer cases. ER beta, on the other hand, is expressed in both the luminal and basal subtypes of breast cancer. The expression of ERs in breast cancer tissue is critical for the success of hormone-based therapies, such as tamoxifen and aromatase inhibitors. These drugs work by blocking oestrogen binding to ERs, thereby slowing down or stopping the cell growth and further cell division of breast cancer (Rai et al., 2002; Wu et al., 2007). The oestrogen-conjugated liposomal preparation of doxorubicin has shown an improvement in doxorubicin uptake by the target cells to treat breast cancer (Rai et al., 2007). *In vivo*, biodistribution confirmed a significantly higher uptake of ooestrogen-conjugated liposome in female albino rats compared to non-conjugated doxorubicin. The aggregation of doxorubicin was found to be 11.05 and 13.9 fold higher compared with the concentration of nontargeted liposomes and free doxorubicin, respectively. Further systemic circulation of drugs was also increased (Rai et al., 2007). Therefore, understanding the biomolecular process that governs the expression of ERs in breast cancer cells is essential for developing targeted therapies that can bypass hormone resistance and get better patient outcomes. Ongoing research is focused on identifying novel biomarkers and signaling pathways that can predict response to hormone-based therapies and promote the practice of personalised medicines in breast cancer cases.

2.5. Integrin Receptors

Integrins have recently been identified as a promising targeting receptor for breast cancer assistance. These glycoproteins lead to cell adhesion and migration, whereas their dysregulation is related to cancer progression and metastasis. Integrins are expressed superficially in tumor tissues and endothelial cells in the cancer microenvironment, making it an appealing therapeutic target site for therapy. One approach to targeting integrins in breast cancer is through the use of liposome formulations. Several studies have demonstrated the potential of liposome-based integrin-targeted therapies for breast cancer (Paliwal et al., 2011). For example, Zhang and his colleagues conducted a study in which a liposomal formulation was targeted to $\alpha v\beta 3$ integrins, an overexpressed receptor present on the surface of breast cancer tissue and the tumour vasculature. The liposomes carried the chemotherapy drug doxorubicin and were shown to effectively retard breast cancer cell proliferation of the breast cancer in a mouse model (Xiong et al., 2005). Similarly, a study by Faezeh and his co-workers developed a liposome formulation that targeted $\alpha 5\beta 1$ integrins, which are also highly expressed in breast cancer tissue. The liposomes were carrying the anticancer drug paclitaxel and were shown to significantly inhibit cell proliferation of breast cancer in a mouse model (Vakhshiteh et al., 2020). In general, these studies suggest that liposome-based integrin-targeted therapies are a promising approach to improve efficacy and reduce side effects suffered during the chemotherapy process. Further exploration is required to optimise these formulations and evaluate their safety and efficacy in clinical trials.

2.6. VEGF Receptors

VEGF receptors or VEGFRs are believed to be the primary protagonist of cancer cell angiogenesis, and their expression level governs the extent of vascularization. Among the VEGFRs, VEGFR-2 is the primary active targeting structure, as it is substantially expressed superficially in the tumour neovasculature and is associated with VEGF binding. Two main strategies for inhibiting angiogenesis have been studied through VEGF and VEGFR-2. In the first strategy, VEGFR-2 is targeted to reduce VEGF binding and trigger an endocytosis pathway, whereas, in the second strategy, VEGF is targeted to prevent ligand binding to VEGFR-2. Work by Waterhouse and colleagues presented a liposomal formulation targeting VEGF receptors to deliver oligonucleotide in malignant cells of breast cancer (Waterhouse et al., 2004). This liposomal preparation substantially decreased the elimination rate from the systemic flow and substantially increased the average area under the curve (AUC) by approximately tenfold in 24 hours. In addition to this approach, anti-VEGFR-2 Mab-labelled liposomes that are 90Y radiolabeled can also be used to target VEGFR-2. The tumour model showed much more delayed tumour growth when exposed to anti-VEGFR-2 Mab-np-90Y compared to 90Y-tagged anti-VEGFR-2 Mab or anti-VEGF-2 Mab alone (Li et al., 2004).

2.7. CD44 Receptors

Hyaluronic acid is an essential element of the extracellular matrix that plays a vital role in various cellular functions, namely growth, migration, and cell division (Sherman et al., 1994). Several hyaluronic acid receptors have been identified, including CD44, Layilin, LYVE-1 and hyaluronic acid-binding proteins present within cells like P-32, IHABP4, and CDC37 (Huang et al., 2000). Overexpression of CD44 receptors is observed in different cancers, including cancer of the stomach, colon, breast and ovary (Herrera et al., 2002). Targeting tumour cells using hyaluronic acid is possible due to its high affinity for the binding and internalisation of hyaluronic acid (Hua et al., 1993). The synthesis of large amounts of hyaluronic acid accelerates cancer proliferation and metastasis. The potential to conjugate with CD44 receptors depends on the size of the hyaluronic acid oligomer (Day et al., 2002). The high specificity and biocompatibility provide the basis for the development of hyaluronic acid-targeted bioconjugates and nano-based delivery systems for anticancer agents. The targeting properties of hyaluronic acid have been assessed by studying doxorubicin-loaded liposomes. The fluidity of the liposome bilayer minimizes the steric interference in the coupling of the ligand with cells having receptors. (Abney et al., 1987). The doxorubicin-loaded liposomes have been prepared by incorporating hyaluronic acid-derivatised phosphatidylethanolamine. These prepared liposomal preparations have shown a greater binding affinity to cells having CD44 receptors (Eliaz et al., 2001).

3. Novel Liposome-Based Methods for Breast Cancer Therapy

Numerous aspects of the cancer microhabitat can be studied to enhance the distribution of nano drug carriers systems. These are enormous prospects, and it is rational to anticipate the fact that they will significantly influence the design of novel emerging eco-friendly drug carriers. In recent years, interest has grown in drug delivery systems that can release active ingredients when triggered by specific stimuli such as ultrasound, electricity, light, temperature, or pH (Figure 2). Furthermore, factors governing liposome-based carriers of anticancer agents and their influence on the management of breast cancer are reviewed in the following section.

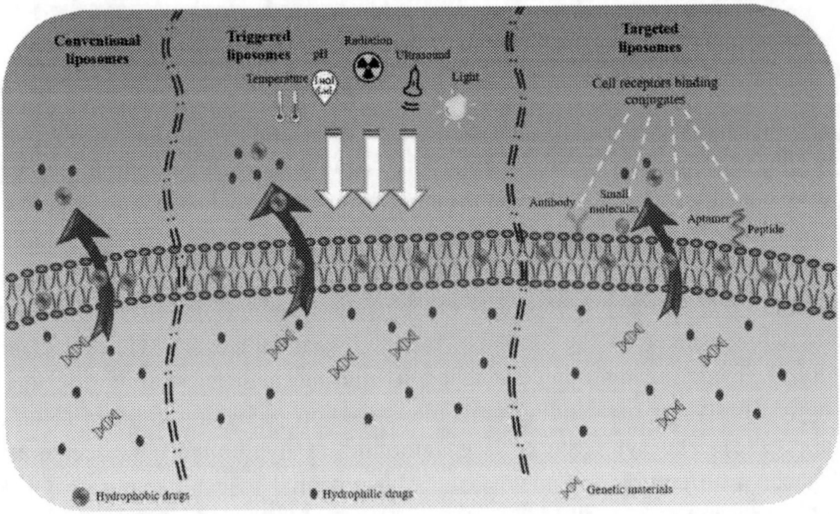

Figure 2. Strategies of liposome modification for targeted liposomal chemotherapy for breast cancer.

3.1. Temperature-Sensitive Liposomes

Temperature-sensitive liposomes can liberate bioactive components that have been enclosed in them when they reach around the temperature of the phase transition (Tm). Transitions of the lipid layer from a gel to a crystalline liquid phase were seen at this Tm (Yatvin et al., 1978). The insertion of lysolipids into the liposomal layer accelerates the release of the encapsulated contents by making it easier to form a transient opening in the bilayer of lipids in Tm (Mills et al., 2004). Hyperthermia is a condition in which the tumour is heated to a temperature between 40°C and 43°C. When administered with the right duration of heat, it facilitates liposome uptake by increasing its concentration two to three times (Leopold et al., 1992). Temperature-sensitive liposomes also aggregate more effectively than traditional liposomes. When exposed to hyperthermia conditions, the liposomes' uptake by the tumour is increased twice as a result of the conjugation of tumor-specific antibodies to liposomes. (Hauck et al., 1997). In this regard, a temperature-sensitive liposomal preparation of doxorubicin, namely, ThermoDox®, has been introduced to treat various cancers, probably breast carcinoma. Intravenous administration of ThermoDox® in conjunction with localised mild hyperthermia heat significantly slows tumour growth by releasing doxorubicin from the liposomal formulation.

3.2. pH-Sensitive Liposomes

pH-responsive liposomes are a type of liposomal system intended to deliver their contents with changes in pH levels. These liposomes can be designed to respond to changes in the interstitial pH levels or the pH levels within the lysosomal / endosomal compartments. The pH levels near tumour cells are typically lower than those found in healthy cells, since there is a formation of lactic acid and ATP hydrolysis in tumour cells (Gerweck et al., 1996). This makes pH-sensitive liposomal delivery systems a promising option to target cancers and deliver anticancer agents. The pH-responsive liposomes are stable in a neutral condition but in acidic conditions i.e., on exposure to lower pH, they become leaky and soft. This allows them to release their contents in response to pathological situations such as infection, inflammation, or tumour (Torchilin et al., 1993). The pH-responsive liposomes are structured to eliminate the issue of recognition and endocytosis by RES. They are made in such a way as to partially release their enclosed active therapeutics substance into the cytosol and rest in the endosome thereby delivering drugs into the cell nucleus or cell cytoplasm (Budker et al., 1996). They are widely used to deliver different kinds of bioactive substances such as anticancer agents, antigens, plasmid DNA, antisense, and oligonucleotides to the cytoplasm *in vitro* (Legendre et al., 1992). pH-triggered delivery of therapeutics to the intracellular region can be achieved by binding them to the cellular level and selective uptake of specific ligand-anchored liposomes. These ligand-anchored liposomes further tend to move toward low pH areas such as lysosomes/endosomes and are then triggered to release the encapsulated drug, thereby achieving drug release through a pH-triggered mechanism (Har-el et al., 2007).

Studies have extensively reported on liposomes with the capability to target anticancer agents in response to the pH of the microenvironment to combat breast cancer. pH-triggered immunoliposomes containing doxorubicin targeted cancer tissues for site-specific delivery of doxorubicin and showed better apoptosis than simple immunoliposomes (Karve et al., 2009). Thus, it can be stated that pH-responsive liposomes have a significant capability to increase the effectiveness of targeted liposomal chemotherapies.

4. Liposome in Gene Therapy for Breast Cancer

An alternative method that could be used to deliver genes is through cationic liposomes. Generally, two methods have been explored for the delivery of genes via liposome; the first method involves the encapsulation of plasmids (Kaneda et al., 1989) and oligonucleotides (Thierry et al., 1992), while in the second method cationic liposome is conjugated with lipoplex (plasmid DNA) (Alio et al., 1996, Crystal et al., 1995). The delivery of cytosolic genes was achieved by activating pH-sensitive liposomes at the target site. The lipoplex fuses with the endosome or plasma membrane and accumulates within the cells (Sharma et al., 1997). The efficacy of transfection of the liposome-DNA complex is highly influenced by the lipids components, particle size and the ratio of DNA: lipids (Zou et al., 2000). Organ and tissue selectivity is one of the major drawbacks of these cationic liposomes; however, this issue can be resolved by modifying cationic liposomes to reduce the adverse effects and increase the delivery of gene to the target cells. Cell-specific targeting ligands can achieve this. Further delivery of genes directly into the cytosol through liposomes can be achieved by using several cell-penetrating peptides, such as transportan, herpex simplex virus type I protein VP22, antennapedia homeodomain and trans-activating transcriptional activator.

To combat breast cancer via gene therapy, the antiangiogenesis approach is commonly used. In this approach, angiostatin or endostatin, which are chemically angiogenic polypeptide inhibitors, are delivered through non-viral vectors (Feldman et al., 2000). Researchers have demonstrated significant inhibition of cell angiogenesis under breast cancer conditions by the conjugate of angiostatin or endostatin encoding plasmid. This discovery laid the groundwork for future advancements in antiangiogenic gene liposomal delivery. In addition to this, the wild-type p53 gene was thoroughly studied to cure breast cancer in nude mice using a liposome-plasmid conjugate. The result showed shrinkage of primary tumor tissue and the prevention of tumor metastasis. However, researchers advise conjugating ligands on liposomes as surface receptors of cells to increase the efficiency of liposome-based gene therapy (Lesoon et al., 1995).

In a study by Xu and his co-workers, a cationic immunoliposome system was developed for the therapy of the p53 cancer suppressor gene circulatory in breast carcinoma in humans. The p53 cancer suppressor gene was tagged with lipid that directed its movement toward breast cancer cells and conjugated with single-chain antibody Fv fragment (scFv) to target transferrin receptors on breast cancer cells. The targeted lipoplex improves the transfection effect, both *in vivo* and *in vitro*, which further causes a significant prolongation of the survival of animal models. Although the expression of the scFv-targeted immunoplex was low in cell culture (Xu et al., 2001). To address this issue, another study was conducted by Xu and his coworkers, and they described a new approach to conjugate the liposome by targeting scFv and different way of expressing anti-transferrin receptor scFv which can produce more protein without being tagged. The liposome system was conjugated with targeting scFv through conjugation of 3 end of cysteine with maleimide component of the liposome. The authors showed that this fusion did not alter the characteristics of scFv such as its ability to target or immunological activity. This preparation could be manufactured at a large scale because of its similarity with that of the recombinant protein. In general, the study highlights the potential of the scFv-targeted immunolipoplex for breast cancer (Xu et al., 2002).

The E1A gene is responsible for producing proteins that hinder HER-2 / neu surface in human breast cancer cells, according to research cited in the reference (Yan et al., 1991). The HER-2 / neu oncogene is highly expressed in different cancers, likely breast cancer in humans. The E1A gene acts as a cancer suppressor gene. They lower the transcription of HER-2/neu and thus suppress cancer growth. However, an effective carrier system is required to deliver a threshold amount of the E1A gene into the targeted breast cancer tissue to achieve desired therapeutic effects. Yoo et al. conducted research on cationic liposomes that were conjugated with plasmid DNA and encode E1A, leading to the composition of a stable lipid E1A complex known as tgDCC-E1A (Yoo et al., 2001). Furthermore, *in vivo* and *in vitro* monitoring showed promising results, leading to clinical trials of liposomal preparation for E1A gene delivery, particularly in patients suffering from recurrent or advanced neck, head, and breast cancers. The outcome of a clinical study conducted in patients with breast cancer showed successful transfer of genes in patients with minimal toxicity when injected with the tgDCC-E1A complex. However, future investigations are necessary to study clinical activity in cancer patients.

5. Liposome Formulation Approved for Breast Cancer Therapy

The first cancer nanomedicines approved by EMA (EU) and FDA (US) was Paclitaxel bounded with albumin marketed under the name Abraxane® and PEGylated liposomal preparation of doxorubicin marketed by the name of Doxil®/Caelyx®. These preparations were associated with the enhancement of the permeability and retention effect of the drug and do not involve drug targeting; therefore they are known as first-generation nanomedicines (Shi et al., 2017; Hare et al., 2017). Currently, only five liposome-based formulations are permitted for practical use to treat breast cancer. Doxil®/Caelyx®, Lipusu®, Lipodox®, Myocet®, Zolsketil® (Table 1).

Caelyx®/Doxil® (commercial name for the same formulation in different countries) is a PEGylated liposome system loaded with Doxorubicin Hydrochloride with a vesicle size of the nanoscale range. It is specifically intended to treat Kaposi's sarcoma related to AIDS, refractory ovarian and metastatic breast cancer, and numerous myeloma. It is the very first nanomedicine to be approved for the clinical application for cancer therapy. This PEGylated liposomal formulation revolutionised cancer therapy as they were able to successfully provide potent anticancer effects with a significant reduction in free doxorubicin levels in the circulating blood while simultaneously increasing the circulation period of anticancer agents

(Barenholz et al., 2012). Later, a critical shortage of Doxil® was suffered in the USA which led to the development of the substrate in the year 2012 and named Lipodox® consisting of a PEGylated liposomal preparation of Doxorubicin hydrochloride (Gaspar et al., 2014).

Table 1. Approved liposomal products marketed for cancer therapy

Clinical Products (Approval year)	Active Agent	Company	Adm. Route	Indication
Doxil® (1995)	Doxorubicin	Sequus Pharmaceuticals	IV	Kaposi sarcoma, ovarian and breast cancer
Caelyx® (1996)	Doxorubicin	Janssen Pharmaceuticals	IV	Recurrent and ovarian breast cancer, Kaposi's sarcoma
DaunoXome® (1996)	Daunorubicin	Galen Limited	IV	Kaposi's sarcoma related to AIDS
Depocyt® (1999)	Cytarabine/ Ara-C	Sigma Tau Pharmaceuticals Inc.	IV	Neoplastic meningitis
Myocet® (2000)	Doxorubicin	Zeneus	IV	Breast cancer
Lipusu® (2006)	Paclitaxel	Luye sike Pharmaceuticals	IV	Ovarian, breast and lung cancer
Mepact® (2009)	Mifamurtide	Takeda France SAS	IV	Non-metastatic osteogenic sarcoma
Marqibo® (2012)	Vincristine	CASI Pharmaceuticals	IV	Lymphoblastic leukaemia (acute)
LipoDox® (2013)	Doxorubicin	Sun Pharma Laboratories Ltd.	IV	Kaposi sarcoma, ovarian and recurrent breast cancer,
Vyxeos® (2017)	Daunorubicin/ Cytarabine	Jazz Pharmaceuticals Inc.	IV	Acute myeloid leukaemia
Zolsketil® (2022)	Doxorubicin	Accord Healthcare	IV	Breast cancer

In contrast, Myocet® is a non-PEGylated liposomal formulation. It contains doxorubicin as an active therapeutic agent and has been used in Europe in the polytherapy of breast carcinoma since 2000. Another drug used in polytherapy along with Myocet® is cyclophosphamide. Myocet® has also been approved for early supervision of metastatic breast carcinoma in the United States of America. An interesting fact is that in the USA, Myocet® was approved due to its capability to minimize treatment-orientated cardiac toxicity instead of better anticancer efficacy (Eloy et al., 2017). Another non-PEGylated liposomal formulation is Lipusu®, which contains paclitaxel as an active pharmaceutical ingredient and is often administered to cure breast and ovarian cancer (Zhu et al., 2019). Lipusu® has given approval for metastatic breast carcinoma in China (AlSawaftah et al., 2021; Wang et al., 2013), but the information on its composition is not publicly available.

Zolsketil® is another PEGylated liposomal preparation approved recently in the year 2022 for the treatment of breast cancer. It contains doxorubicin hydrochloride with methoxypolyethylene glycol that is bound to the surface of the liposome. This process protects liposomes from being phagocytes by the body's immune system (mononuclear phagocyte system), which ultimately increases the interval of liposomal preparation in the bloodstream. Zolsketil® is used as a monotherapy for breast cancer remedies, specifically in those cases where the risk is very high. It is also used in the remedy for advanced ovarian cancer in patients whose platinum-based first-line chemotherapy has failed to achieve desired results and treatment of Kaposi' sarcoma related to AIDS.

The liposomal formulations prepared so far rely on their size and preferably accumulation of active pharmaceutical ingredients in the interstitial spaces of cancer cells through enhanced permeability and retention effects through passive targeting. Formulation based on active targeting systems is still the hotspot of research. Ongoing clinical trials on liposomal preparation to treat breast cancer are listed in Table 1.

6. Liposomes in the Treatment of Cervical Cancer

Cervical carcinoma is a form of cancer that affects the cervix, which is the lower part of the uterus. It is one of the most common forms of cancer in females worldwide and is caused by human papillomavirus (HPV). Cervical carcinoma is a major health concern for women worldwide, and traditional chemotherapy treatments can have significant side effects and may not be effective in all cases. They are nonspecific in nature and suffer from the issue of drug resistance. Liposomes have shown promising results as a potential treatment option for cervical carcinoma. Many studies have investigated the use of liposomes in the treatment of cervical carcinoma. An efficient first-line chemotherapy medicine for cervical cancer is cisplatin. Using liposomes with polylactic-coglycolic acids (PLGA), Dana et al. created a cisplatin-encapsulated carrier system that has successfully minimised drug toxicity and resistance. An antiangiogenic drug, Avastin was further linked with the lipid system using the double emulsion solvent evaporation approach. This Avastin conjugated system improved the potential for cellular uptake and increased binding ability in 3D spheroid and xenograft studies (Dana et al., 2020).

Another team created a special type of conjugate that joined CD59 and miRNA-1284 with cisplatin and then linked liposomes to it. Compared to either cisplatin or miR-1284 alone, this co-delivery approach has shown significantly more anticancer property in cervical cancer cells, and the number of cell death was much higher (60 vs. 20% and 12%, respectively) (Wang et al., 2020). A novel anticancer treatment called photodynamic therapy has drawn much attention lately. Reactive oxygen species and free radicals can be released by photodynamic therapy at the appropriate absorption wavelength, which is entirely dependent on photosensitisers (Fang et al., 2020). Estrone-targeted PEGylated cisplatin was created and characterized using a thin-film hydration process. With a surface charge, 97.3 nm particle size, and an encapsulation effectiveness, cisplatin liposomal preparation demonstrated a spherical structure. According to an *in vivo* targeting evaluation, the cisplatin liposomal preparation could specifically accumulate at the cancer site of mice carrying HeLa. Using the MTT assay, cytotoxicity tests in HeLa cells indicated that cisplatin liposomal preparation had a greater cytotoxic effect. The most effective tumour inhibition was observed in HeLa-negative mice when cisplatin liposomal preparation was used as an *in vivo* anti-tumor agent. The pharmacokinetics and biodistribution demonstrated a better metabolism of cisplatin. The acute toxicity showed that in healthy ICR mice, cisplatin liposomal preparation could increase LD50 and lower myelosuppression (Li et al., 2022).

In addition, the side effect and efficacy of the paclitaxel-encapsulated liposomal formulation was done for neoadjuvant chemotherapy in locally spread cervical cancer at an advanced stage. Paclitaxel liposome can achieve efficacy comparable to standard paclitaxel in paclitaxel-platinum neoadjuvant chemotherapy of locally advanced cervical cancer, but also helps to lower neurotoxicity, alopecia, and myalgia (Wang et al., 2019). The con-encapsulated liposomal system of cisplatin and mifepristone significantly reduced tumour size both *in vivo* and *in vitro,* while causing no systemic damage in the experimental animals. Apoptosis was up and there was evidence of cell cycle arrest. However, further studies are necessary in light of the encouraging outcomes of coencapsulation of cisplatin and mifepristone in liposomes for the management of cervical cancer (Ledezma et al., 2020). By substituting platinum-based medications, ruthenium complexes have expanded the potential for advancement in current cervical cancer treatment. This ruthenium can be complexed with curcumin, a biocompatible flavonoid with immense therapeutic effects, making it beneficial in reducing metal-related toxicity. The liposomal preparation containing ruthenium (II) and curcumin prepared by the thin layer evaporation method has significant cytotoxicity in human cervical cancer cell lines (HeLa) (Lakshmi et al., 2021). Lymoquinone-loaded liposomal drug delivery systems also showed better results in cell line analyses compared to thymoquinone alone, and this finding may have a significant impact on how breast and cervical cancer patients are treated (Shariare et al., 2022).

The targeted approach of liposomes makes them an attractive option for delivering drugs and other therapies directly to cancer cells, potentially reducing side effects and increasing effectiveness. However, more research is required to fully comprehend the potential of liposomes to treat cervical carcinoma and other cancers.

7. Liposome in the Treatment of Ovarian Cancer

Ovarian cancer is an aggressive malignancy that affects women all over the world. It is often diagnosed at a later stage, making it difficult to treat and resulting in a poor prognosis. Traditional chemotherapy approaches have limited efficacy due to inheritance of drug resistance and adverse effects. Therefore, there is an immediate need to discover new strategies for ovarian cancer treatment that can improve patient outcomes. One such approach with significant potential is to use liposomes as drug carriers. Liposomes offer several benefits in drug delivery, including increased bioavailability, prolonged circulation time, targeted drug delivery, and reduced toxicity compared to conventional chemotherapeutic approaches (Shah et al., 2018). Many preclinical and clinical data have illustrated the potential of liposomal formulations in the treatment of ovarian cancer.

For instantaneous, liposomal preparation of a vinca alkaloid-based Prodrug, CPD100, was prepared. With an aqueous solubility of 33 mg / ml and molecular weight of 900 g/mole, the hydrophilic molecule CPD100 is perfectly suitable for the trapping of the drug. The recommended method for liposomal encapsulation is active drug loading, which fully depends on the physicochemical properties. The evaluation of CPD100Li *in vitro* involves stability evaluation for freshly manufactured and freeze-dried CPD100Li, as well as loading and size characterisation. Using a comparison of the mechanism of action and the effect of CPD100 on vinblastine in hypoxic and normoxic circumstances, as well as the impact on cell proliferation, cell division, and apoptosis, CPD100Li is evaluated in ovarian cancer cell lines. Pharmacokinetic, maximum tolerated dose (MTD) and dose-limiting toxicity (DLT) tests are all part of the *in vivo* characterisation of CPD100Li. The evaluation of CPD100Li *in vivo* also establishes the requirement to add the ionophore to improve stability of the liposome. This clinical evaluation is significant since the FDA has not yet classified ionophore A23187 as generally regarded safe (GRAS), even though it has been shown to be an essential component to retain anticancer drugs in liposomes. A crucial stage in further clinical research is determining the pharmacokinetics, DLT, and MTD of the ionophore containing CPD100Li (Shah et al., 2018). Similarly, liposomal paclitaxel (LEP-ETU) has been shown to have a better efficacy with negligible toxicity than conventional paclitaxel in preclinical studies (Zhang et al., 2005).

Currently, only three liposomal formulations are approved for clinical application in the management of breast cancer: Doxil®/Caelyx®, Lipusu®, and Lipodox® (Table 1). Caelyx® /Doxil® (commercial name in US and Europe, respectively), a PEGylated liposomal drug carrier loaded with doxorubicin, has shown promising results in phase III clinical trials for recurrent ovarian cancer (Barenholz et al., 2012). Another PEGylated liposomal formulation, namely Lipodox® loaded with doxorubicin hydrochloride, has been prepared after the shortage of Doxil® in the USA in 2012 (Barenholz et al., 2012). Furthermore, a non-PEGylated liposomal formulation, Lipusu® encapsulating paclitaxel as an active pharmaceutical ingredient, is often administered for the treatment of breast and ovarian cancer (Gaspar et al., 2014).

Moreover, targeted liposomes that incorporate tumour-specific ligands on their surface exhibit enhanced tumour accumulation, selectivity, and efficacy. For example, a liposomal formulation targeting the folate receptor was loaded with carboplatin and has been shown to successfully target the ovarian cancer cells and further kill them, both *in vivo* and *in vitro* (Tang et al., 2022). The formulation of the AS1411 Aptamer-functionalised liposome revealed aptamer improved the targeting of the carrier through the nucleolin-mediated transmembrane phagocytosis process in tissue of ovarian cancer. In ovarian cancer tissue, miR-29b, based on lipofectamine, showed significant mortality depending upon the concentration. The vitality of A2780 cells was decreased by LP-miR when compared to the control group (group untreated), but was not affected by mutant loading, highlighting the importance of accurate gene sequencing. In contrast to the untreated control group, cells treated with LP-miR showed a significant increase in Annexin-V+ and PI+ cells, demonstrating the strength of miR-29b. This new liposome containing miR-29b and directed by aptamer may provide a new background for research and has the potential to be a breakthrough treatment platform for ovarian malignancies (Chaudhury et al., 2012).

In general, liposomes can revolutionize the treatment of ovarian cancer by adding more effective, targeted, and less toxic therapeutic measures for cancer patients. Although the clinical application of liposomes in ovarian cancer management is still in its early stages, the promising preclinical and clinical results imply that liposomes could become a powerful tool to treat ovarian cancer. Ongoing clinical trials on liposomal preparation for the treatment of ovarian cancer are listed in Table 2.

8. Challenges and Future Directions for Liposomal Preparation

With the emergence of nanotechnology, the use of liposome systems as carriers of anticancer agents in cancer treatment techniques and methods is increasing broadly. Some lipid preparations have been successfully marketed. After so many developments, there are some challenges in the transformation process and thin clinical use. Liposomal stability is one of them which governs changes in loading, release rate, and permeability of therapeutic drugs during their preparation, storage, and further metabolism of encapsulated drugs. The physical and chemical stability of liposomal preparation regulates the marketing and clinical advantages of liposomes as therapeutic drug delivery systems. The membrane of liposomes is a phospholipid in nature; these phospholipids can allow liposomal particles to aggregate. Sedimentation and formation of unstable phospholipid layers due to continuous transmembranous movement lead to instability of the morphological and physical structure of liposomes, while hydrolysis and oxidation are the major reason for the chemical instability of liposomes (Sercombe et al., 2012).

Table 2. Ongoing clinical trial on liposomal preparation for the treatment of breast cancer and ovarian cancer

NCT Number	Objective	Cancer type	Drugs	Phase	Dose	Sponsor/Collaborators
NCT05273944	Bioequivalence Study	Breast Cancer, Ovarian Cancer	Liposomal doxorubicin hydrochloride	Phase 1	50 mg/m^2	Shenzhen Kangzhe Pharmaceutical Co. Ltd., 3 universities and 8 hospitals
NCT04244552	Evaluate pharmacokinetics, biological activity, safety, and tolerability	Breast Cancer, Ovarian Cancer and 14 more	ATRC-101 + pembrolizumab + Pegylated liposomal doxorubicin	Phase 1	40 mg/m^2	Atreca Inc.
NCT04718376	Evaluate the safety and efficacy	Platinum resistant Ovarian Cancer	Liposomal mitoxantrone hydrochloride	Phase 1	20 mg/m^2	CSPC ZhongQi pharmaceutical Tech. Co., Ltd
NCT05261490	Evaluate the safety and efficacy of combination therapy	Ovarian Cancer, Fallopian tube cancer	Maplirpacept + Pegylated liposomal doxorubicin	Phase 1 Phase 2	40 mg/m^2	Pfizer
NCT05386524	Evaluate the safety and efficacy of combination therapy	Breast Cancer	Sintilimab + biosimilar bevacizumab + pegylated liposomal doxorubicin	Phase 2	30 mg/m^2	Fudan University
NCT05346107	Evaluate the safety and efficacy	breast cancer, HER2-positive breast cancer	Nab-paclitaxel + PEGylated liposomal doxorubicin + cyclophosphamide	Phase 2	35 mg/m^2	Xijing hospital
NCT04927481	Evaluate the safety and efficacy	Breast Cancer	Liposomal mitoxantrone hydrochloride	Phase 2	20 mg/m^2	CSPC ZhongQi Pharmaceutical Tech. Co. Ltd.
NCT04872985	Evaluate the safety and efficacy	breast cancer, hormone-receptor positive breast cancer	Liposomal doxorubicin hydrochloride	Phase 2	30 mg/m^2	Jiangsu HengRui Medicine Co. Ltd., Sun Yat-sen University

Table 2. (Continued)

NCT Number	Objective	Cancer type	Drugs	Phase	Dose	Sponsor/Collaborators
NCT04603911	Investigate the relative efficacy	Breast Cancer	Liposomal bupivacaine + Bupivacaine, epinephrine, dexamethasone, and clonidine	Phase 2	-	Tufts Medical Center
NCT03971409	Investigate the relative efficacy	Stage III, IIIA, IIIB, IIIC, IV breast cancer, triple negative, non-resectable, invasive and recurrent breast cancer	Sacituzumab govitecan + avelumab + binimetinib + utomilumab + liposomal doxorubicin	Phase 2	10 mg/kg	Pfizer, Johns Hopkins University, and 6 more
NCT03409198	Evaluation of toxicity of combined treatment	Metastatic breast cancer, breast cancer	Ipilimumab + Nivolumab + cyclophosphamide + pegylated liposomal doxorubicin	Phase 2	20 mg/m^2	Oslo University Hospital, Helse Stavanger HF and 6 more
NCT03164993	Evaluate the safety and efficacy	Triple-negative breast cancer, breast cancer	Atezolizumab + Pegylated liposomal doxorubicin + cyclophosphamide	Phase 2	20 mg/m^2	Oslo University Hospital, Norwegian Cancer Society and 8 more
NCT03071926	Evaluate the safety and efficacy	Breast cancer	PEGylated liposomal doxorubicin	Phase 2	20 mg	Fudan University
NCT02456857	Evaluate the efficacy	Invasive breast carcinoma, Triple-negative breast cancer	Everolimus + Bevacizumab + PEGylated liposomal doxorubicin hydrochloride	Phase 2	-	M.D. Anderson Cancer Center
NCT02315196	Evaluate the relative efficacy	Triple-negative, HER2-negative, oestrogen and progesterone receptor-negative breast cancer, Stage IIA, IIB, IIIA, IIIB, IIIC Breast Cancer	Epirubicin hydrochloride + paclitaxel + carboplatin + PEGylated liposomal doxorubicin hydrochloride	Phase 2	-	National Cancer Institute, Rutgers Cancer Institute of New Jersey

NCT Number	Objective	Cancer type	Drugs	Phase	Dose	Sponsor/Collaborators
NCT01210768	Evaluate safety, overall survival and quality of life	Breast Cancer	Epirubicin + cyclophosphamide + liposomal-doxorubicin	Phase 2	37.5 mg/m^2	TTY Biopharm
NCT05159193	Evaluate the safety and efficacy	Breast Cancer	pegylated liposomal doxorubicin + cyclophosphamid+ trastuzumab + pertuzumab + docetaxel + carboplatin	Phase 3	30 mg/m^2	Sun Yat-Sen University, CSPC Ouyi Pharmaceutical Co. Ltd.
NCT00326456	Evaluate the relative efficacy, quality of life, overall survival and toxicity	Ovarian Cancer	Liposomal doxorubicin + Paclitaxel + Carboplatin	Phase 3	30 mg/m^2	National Cancer Institute, Naples
NCT04849858	Improve pain control and lower risk of post-operative side effects	Uterine cancer and Ovarian cancer	bupivacaine + Liposomal bupivacaine	Phase 3	-	University of California, Irvine
NCT05302336	Evaluate the safety and efficacy	Breast cancer and Ovarian cancer	Liposomal doxorubicin + cyclophosphamide	Phase 4	35 mg/m^2	School of Medicine, Zhejiang University

The selection of an appropriate method of sterilisation while maintaining the stability of the liposomal preparation is a great challenge during manufacturing since the majority of liposomal preparations are intended for parenteral administration. Today, sterilization is done by steam filtration and radiation sterilisation via gamma rays. Liposomes are sensitive to any chemical or/and physical deterioration system; they can change the physiochemical properties of the liposomal component and increase the toxicity of the final product when exposed to heat, radiation, and toxic chemicals. Furthermore, clogging of liposomal membrane, retention of small bacteria, and loss of integrity of liposomes are other issues that can occur (Delma et al., 2021). Further improvements are required in the liposome system to improve its therapeutic performance. Novel targeting techniques, gene mutations, and smart materials are likely to help increase the survival rate of cancer patients. Another way to provide a more successful targeted treatment is the co-administration of various drugs and non-coding RNAs (siRNAs) via liposomes. Targeting agents and stimuli-responsive molecules can be combined via liposomes to produce controlled delivery and release of therapeutic moieties.

Currently, research has not fully explored the previously combinatorial methods used in the chemotherapy of breast cancer, cervical cancer and ovarian cancer. Additionally, *in vivo* research at various clinical stages is required to study the efficacy of novel liposomal formulations tested in the laboratory. This is necessary to estimate the safety and efficacy of liposomal preparation to cure cancer. Also, it is important for surface engineered liposomes loaded with either drug or gene to be reproducible from batch-to-batch production if they are going to be part of the fate of clinical carcinoma therapeutics. This means that each batch must be consistent in order to ensure that the treatment is effective and reliable.

References

Abney JR, Braun J, Owicki JC. Lateral interactions among membrane proteins. Implications for the organization of gap junctions. *Biophys J.* 1987 Sep;52(3):441–54.

Ajeeshkumar KK, Aneesh PA, Raju N, Suseela M, Ravishankar CN, Benjakul S. Advancements in liposome technology: Preparation techniques and applications in food, functional foods, and bioactive delivery: A review. *Compr Rev Food Sci Food Saf.* 2021 Mar 1;20(2):1280–306.

Aliño SF, Bobadilla M, Crespo J, Lejarreta M. Humanα1-Antitrypsin Gene Transfer to*In Vivo*Mouse Hepatocytes. *Hum Gene Ther.* 1996;7(4):531–6.

Allred DC, Mohsin SK, Fuqua SAW. Histological and biological evolution of human premalignant breast disease. *Endocr Relat Cancer.* 2001;8(1):47–61.

AlSawaftah N, Pitt WG, Husseini GA. Dual-Targeting and Stimuli-Triggered Liposomal Drug Delivery in Cancer Treatment. *ACS Pharmacol Transl Sci.* 2021 Jun 11;4(3):1028–49.

Ansari L, Shiehzadeh F, Taherzadeh Z, Nikoofal-Sahlabadi S, Momtazi-borojeni AA, Sahebkar A, et al. The most prevalent side effects of pegylated liposomal doxorubicin monotherapy in women with metastatic breast cancer: a systematic review of clinical trials. *Cancer Gene Ther.* 2017;24(5):189–93.

Antoniou AI, Giofrè S, Seneci P, Passarella D, Pellegrino S. Stimulus-responsive liposomes for biomedical applications. *Drug Discov Today*. 2021;26(8):1794–824.

Arab-Tehrany E, Elkhoury K, Francius G, Jierry L, Mano JF, Kahn C, et al. Curcumin Loaded Nanoliposomes Localization by Nanoscale Characterization. Vol. 21, *International Journal of Molecular Sciences*. 2020.

Australian Institute of Health and Welfare. Cancer in Australia. *Cancer series* no. 133. Cat no. CAN 144. Canberra:AIHW; 2021.

Badran MM, Alouny NN, Aldosari BN, Alhusaini AM, Abou El Ela AES. Transdermal Glipizide Delivery System Based on Chitosan-Coated Deformable Liposomes: Development, Ex Vivo, and *In Vivo* Studies. *Pharmaceutics*. 2022 Apr;14(4).

Bangham AD, Horne RW. Negative staining of phospholipids and their structural modification by surface-active agents as observed in the electron microscope. *J Mol Biol*. 1964 Jan 1;8(5):660-IN10.

Barenholz Y (Chezy). Doxil® — The first FDA-approved nano-drug: Lessons learned. *J Control Release*. 2012;160(2):117–34.

Bernier-Latmani J, Petrova T V. Intestinal lymphatic vasculature: structure, mechanisms and functions. *Nat Rev Gastroenterol Hepatol*. 2017 Sep;14(9):510–26.

Bilal M, Qindeel M, Raza A, Mehmood S, Rahdar A. Stimuli-responsive nanoliposomes as prospective nanocarriers for targeted drug delivery. *J Drug Deliv Sci Technol.* 2021;66:102916.

Budker V, Gurevich V, Hagstrom JE, Bortzov F, Wolff JA. pH-sensitive, cationic liposomes: A new synthetic virus-like vector. *Nat Biotechnol*. 1996;14(6):760–4.

Cardoso F, Spence D, Mertz S, Corneliussen-James D, Sabelko K, Gralow J, et al. Global analysis of advanced/metastatic breast cancer: Decade report (2005–2015). *Breast.* 2018 Jun 1;39:131–8.

Chaudhury A, Das S, Bunte RM, Chiu GNC. Potent therapeutic activity of folate receptor-targeted liposomal carboplatin in the localized treatment of intraperitoneally grown human ovarian tumor xenograft. *Int J Nanomedicine*. 2012;7:739–51.

Crystal RG. Transfer of Genes to Humans: Early Lessons and Obstacles to Success. *Science* (80-). 1995;270(5235):404–10.

Dana P, Bunthot S, Suktham K, Surassmo S, Yata T, Namdee K, et al. Active targeting liposome-PLGA composite for cisplatin delivery against cervical cancer. *Colloids Surfaces B Biointerfaces*. 2020;196:111270.

Day AJ, Prestwich GD. Hyaluronan-binding Proteins: Tying Up the Giant. *J Biol Chem.* 2002;277(7):4585–8.

Delma KL, Lechanteur A, Evrard B, Semdé R, Piel G. Sterilization methods of liposomes: Drawbacks of conventional methods and perspectives. *Int J Pharm.* 2021 Mar;597:120271.

Di paolo A. Liposomal Anticancer Therapy: Pharmacokinetic and Clinical Aspects. *J Chemother*. 2004 Nov 1;16(sup4):90–3.

Eliaz RE, Szoka FCJ. Liposome-encapsulated doxorubicin targeted to CD44: a strategy to kill CD44-overexpressing tumor cells. *Cancer Res.* 2001 Mar;61(6):2592–601.

Elkhoury K, Sanchez-Gonzalez L, Lavrador P, Almeida R, Gaspar V, Kahn C, et al. Gelatin Methacryloyl (GelMA) *Nanocomposite Hydrogels Embedding Bioactive Naringin Liposomes*. Vol. 12, Polymers. 2020.

Eloy JO, Claro de Souza M, Petrilli R, Barcellos JPA, Lee RJ, Marchetti JM. Liposomes as carriers of hydrophilic small molecule drugs: Strategies to enhance encapsulation and delivery. *Colloids Surfaces B Biointerfaces*. 2014;123:345–63.

Eloy JO, Petrilli R, Trevizan LNF, Chorilli M. Immunoliposomes: A review on functionalization strategies and targets for drug delivery. *Colloids Surfaces B Biointerfaces*. 2017;159:454–67.

Fang X, Xie A, Song H, Jiang D, Li H, Wang Z, et al. A novel α-(8-quinolinyloxy) monosubstituted zinc phthalocyanine nanosuspension for potential enhanced photodynamic therapy. *Drug Dev Ind Pharm.* 2020 Nov;46(11):1881–8.

Feldman AL, Libutti SK. Progress in antiangiogenic gene therapy of cancer. *Cancer*. 2000;89(6):1181–94.

Ferguson LT, Ma X, Myerson JW, Wu J, Glassman PM, Zamora ME, et al. Mechanisms by Which Liposomes Improve Inhaled Drug Delivery for Alveolar Diseases. *Adv NanoBiomed Res*. 2023;3(3):2200106.

Gaspar RS, Florindo HF, Silva LC, Videira MA, Corvo ML, Martins BF, et al. *Regulatory Aspects of Oncologicals: Nanosystems Main Challenges BT - Nano-Oncologicals: New Targeting and Delivery Approaches.* In: Alonso MJ, Garcia-Fuentes M, editors. Cham: Springer International Publishing; 2014. p. 425–52.

Gerweck LE, Seetharaman K. Cellular pH gradient in tumor versus normal tissue: potential exploitation for the treatment of cancer. *Cancer Res.* 1996 Mar;56(6):1194–8.

Goyal P, Goyal K, Vijaya Kumar SG, Singh A, Katare OP, Mishra DN. Liposomal drug delivery systems--clinical applications. *Acta Pharm.* 2005 Mar;55(1):1–25.

Gustafson HH, Holt-Casper D, Grainger DW, Ghandehari H. Nanoparticle Uptake: The Phagocyte Problem. *Nano Today*. 2015 Aug;10(4):487–510.

Han Y, Gao Z, Chen L, Kang L, Huang W, Jin M, et al. Multifunctional oral delivery systems for enhanced bioavailability of therapeutic peptides/proteins. *Acta Pharm Sin B*. 2019 Sep;9(5):902–22.

Harashima H, Sakata K, Funato K, Kiwada H. Enhanced hepatic uptake of liposomes through complement activation depending on the size of liposomes. *Pharm Res*. 1994 Mar;11(3):402–6.

Hare JI, Lammers T, Ashford MB, Puri S, Storm G, Barry ST. Challenges and strategies in anti-cancer nanomedicine development: An industry perspective. *Adv Drug Deliv Rev*. 2017;108:25–38.

Har-el Y, Kato Y. Intracellular Delivery of Nanocarriers for Cancer Therapy. *Curr Nanosci*. 2007;3(4):329–38.

Hauck ML, Coffin DO, Dodge RK, Dewhirst MW, Mitchell JB, Zalutsky MR. A local hyperthermia treatment which enhances antibody uptake in a glioma xenograft model does not affect tumour interstitial fluid pressure. *Int J Hyperth*. 1997;13(3):307–16.

Herrera-Gayol A, Jothy S. Effect of Hyaluronan on Xenotransplanted Breast Cancer. *Exp Mol Pathol*. 2002;72(3):179–85.

Hua Q, Knudson CB, Knudson W. Internalization of hyaluronan by chondrocytes occurs via receptor-mediated endocytosis. *J Cell Sci.* 1993;106(1):365–75.

Huang S, New L, Pan Z, Han J, Nemerow GR. Urokinase Plasminogen Activator/Urokinase-specific Surface Receptor Expression and Matrix Invasion by Breast Cancer Cells Requires Constitutive p38α Mitogen-activated Protein Kinase Activity. *J Biol Chem*. 2000;275(16):12266–72.

Huckaby JT, Lai SK. PEGylation for enhancing nanoparticle diffusion in mucus. *Adv Drug Deliv Rev*. 2018 Jan;124:125–39.

Hung MC, Matin A, Zhang Y, Xing X, Sorgi F, Huang L, et al. HER-2/neu-targeting gene therapy--a review. *Gene*. 1995 Jun;159(1):65–71.

Kaneda Y, Iwai K, Uchida T. Increased Expression of DNA Cointroduced with Nuclear Protein in Adult Rat Liver. *Science* (80-). 1989;243(4889):375–8.

Kanwal U, Irfan Bukhari N, Ovais M, Abass N, Hussain K, Raza A. Advances in nano-delivery systems for doxorubicin: an updated insight. *J Drug Target*. 2018 Apr 21;26(4):296–310.

Kapoor MS, D'Souza A, Aibani N, Nair SS, Sandbhor P, Kumari D, et al. Stable Liposome in Cosmetic Platforms for Transdermal Folic acid delivery for fortification and treatment of micronutrient deficiencies. *Sci Rep*. 2018 Oct 31;8(1):16122.

Karve S, Alaouie A, Zhou Y, Rotolo J, Sofou S. The use of pH-triggered leaky heterogeneities on rigid lipid bilayers to improve intracellular trafficking and therapeutic potential of targeted liposomal immunochemotherapy. *Biomaterials*. 2009;30(30):6055–64.

Kurano T, Kanazawa T, Ooba A, Masuyama Y, Maruhana N, Yamada M, et al. Nose-to-brain/spinal cord delivery kinetics of liposomes with different surface properties. *J Control Release*. 2022;344:225–34.

Lakshmi BA, Reddy AS, Sangubotla R, Hong JW, Kim S. Ruthenium(II)-curcumin liposome nanoparticles: Synthesis, characterization, and their effects against cervical cancer. *Colloids Surf B Biointerfaces.* 2021 Aug;204:111773.

Lasic DD, Templeton NS. Liposomes in gene therapy. *Adv Drug Deliv Rev.* 1996;20(2–3):221–66.

Lasic DD. Novel applications of liposomes. *Trends Biotechnol.* 1998;16(7):307–21.

Ledezma-Gallegos F, Jurado R, Mir R, Medina LA, Mondragon-Fuentes L, Garcia-Lopez P. Liposomes Co-Encapsulating Cisplatin/Mifepristone Improve the Effect on Cervical Cancer: *In Vitro* and *In Vivo* Assessment. *Pharmaceutics*. 2020 Sep;12(9).

Legendre JY, Szoka FCJ. Delivery of plasmid DNA into mammalian cell lines using pH-sensitive liposomes: comparison with cationic liposomes. *Pharm Res.* 1992 Oct;9(10):1235–42.

Leopold KA, Dewhirst M, Samulski T, Harrelson J, Tucker JA, George SL, et al. Relationships among tumor temperature, treatment time, and histopathological outcome using preoperative hyperthermia with radiation in soft tissue sarcomas. *Int J Radiat Oncol*. 1992;22(5):989–98.

Lesoon-Wood LA, Kim WH, Kleinman HK, Weintraub BD, Mixson AJ. Systemic Gene Therapy with p53 Reduces Growth and Metastases of a Malignant Human Breast Cancer in Nude Mice. *Hum Gene Ther.* 1995;6(4):395–405.

Li L, Wartchow CA, Danthi SN, Shen Z, Dechene N, Pease J, et al. A novel antiangiogenesis therapy using an integrin antagonist or anti-Flk-1 antibody coated 90Y-labeled nanoparticles. *Int J Radiat Oncol Biol Phys*. 2004 Mar;58(4):1215–27.

Li Q, Zhu M, Li Y, Tang H, Wang Z, Zhang Y, et al. Estrone-targeted PEGylated Liposomal Nanoparticles for Cisplatin (DDP) Delivery in Cervical Cancer. *Eur J Pharm Sci Off J Eur Fed Pharm Sci*. 2022 Jul;174:106187.

Li Y-J, Lei Y-H, Yao N, Wang C-R, Hu N, Ye W-C, et al. Autophagy and multidrug resistance in cancer. *Chin J Cancer*. 2017;36(1):52.

Miele E, Spinelli GP, Miele E, Tomao F, Tomao S. Albumin-bound formulation of paclitaxel (Abraxane® ABI-007) in the treatment of breast cancer. *Int J Nanomedicine*. 2009;4(1):99–105.

Mills JK, Needham D. The Materials Engineering of Temperature-Sensitive Liposomes. *Methods in Enzymology*. Elsevier; 2004. p. 82–113.

Mirtaleb MS, Shahraky MK, Ekrami E, Mirtaleb A. Advances in biological nano-phospholipid vesicles for transdermal delivery: A review on applications. *J Drug Deliv Sci Technol*. 2021;61:102331.

Mokhtari-Hessari P, Montazeri A. Health-related quality of life in breast cancer patients: review of reviews from 2008 to 2018. *Health Qual Life Outcomes*. 2020 Oct;18(1):338.

Muggia FM. Doxorubicin-polymer conjugates: further demonstration of the concept of enhanced permeability and retention. Vol. 5, *Clinical cancer research : an official journal of the American Association for Cancer Research. United States*; 1999. p. 7–8.

Nsairat H, Khater D, Sayed U, Odeh F, Al Bawab A, Alshaer W. Liposomes: structure, composition, types, and clinical applications. *Heliyon*. 2022 May;8(5):e09394.

Paliwal SR, Paliwal R, Agrawal GP, Vyas SP. Liposomal nanomedicine for breast cancer therapy. *Nanomedicine*. 2011;6(6):1085–100.

Park JW, Kirpotin DB, Hong K, Shalaby R, Shao Y, Nielsen UB, et al. Tumor targeting using anti-her2 immunoliposomes. *J Control release Off J Control Release Soc.* 2001 Jul;74(1–3):95–113.

Press D. *Drug Delivery approaches for breast cancer*. 2017;6205–18.

Rai S, Paliwal R, Vaidya B, Gupta PN, Mahor S, Khatri K, et al. Estrogen(s) and analogs as a non-immunogenic endogenous ligand in targeted drug/DNA delivery. *Curr Med Chem*. 2007;14(19):2095–109.

Rait AS, Pirollo KF, Xiang L, Ulick D, Chang EH. Tumor-targeting, systemically delivered antisense HER-2 chemosensitizes human breast cancer xenografts irrespective of HER-2 levels. *Mol Med.* 2002 Aug;8(8):475–86.

Sercombe L, Veerati T, Moheimani F, Wu SY, Sood AK, Hua S. Advances and Challenges of Liposome Assisted Drug Delivery. *Front Pharmacol.* 2015;6:286.

Shah VM, Nguyen DX, Al Fatease A, Patel P, Cote B, Woo Y, et al. Liposomal formulation of hypoxia activated prodrug for the treatment of ovarian cancer. *J Control Release*. 2018;291:169–83.

Shariare MH, Khan MA, Al-Masum A, Khan JH, Uddin J, Kazi M. Development of Stable Liposomal Drug Delivery System of Thymoquinone and Its *In Vitro* Anticancer Studies Using Breast Cancer and Cervical Cancer Cell Lines. *Molecules.* 2022 Oct;27(19).

Sharma A. Liposomes in drug delivery: Progress and limitations. *Int J Pharm.* 1997;154(2):123–40.

Sherman L, Sleeman J, Herrlich P, Ponta H. Hyaluronate receptors: key players in growth, differentiation, migration and tumor progression. *Curr Opin Cell Biol.* 1994;6(5):726–33.

Shi J, Kantoff PW, Wooster R, Farokhzad OC. Cancer nanomedicine: progress, challenges and opportunities. *Nat Rev Cancer.* 2017;17(1):20–37.

Stroock AD, Dertinger SKW, Ajdari A, Mezic I, Stone HA, Whitesides GM. Chaotic mixer for microchannels. *Science*. 2002 Jan;295(5555):647–51.

Sung H, Ferlay J, Siegel RL, Laversanne M, Soerjomataram I, Jemal A, et al. Global Cancer Statistics 2020: GLOBOCAN Estimates of Incidence and Mortality Worldwide for 36 Cancers in 185 Countries. *CA Cancer J Clin.* 2021 May 1;71(3):209–49.

Taha EI, El-Anazi MH, El-Bagory IM, Bayomi MA. Design of liposomal colloidal systems for ocular delivery of ciprofloxacin. *Saudi Pharm J SPJ Off Publ Saudi Pharm Soc*. 2014 Jul;22(3):231–9.

Tang H, Xie Y, Zhu M, Jia J, Liu R, Shen Y, et al. Estrone-Conjugated PEGylated Liposome Co-Loaded Paclitaxel and Carboplatin Improve Anti-Tumor Efficacy in Ovarian Cancer and Reduce Acute Toxicity of Chemo-Drugs. *Int J Nanomedicine*. 2022;17:3013–41.

Thierry AR, Dritschilo A. Intracellular availability of unmodified, phosphorothioated and liposomally encapsulated oligodeoxynucleotides for antisense activity. *Nucleic Acids Res.* 1992 Nov 11;20(21):5691–8.

Tonglairoum P, Brannigan RP, Opanasopit P, Khutoryanskiy V V. Maleimide-bearing nanogels as novel mucoadhesive materials for drug delivery. *J Mater Chem B*. 2016 Oct;4(40):6581–7.

Torchilin VP, Zhou F, Huang L. pH-Sensitive Liposomes. *J Liposome Res.* 1993;3(2):201–55.

Vakhshiteh F, Khabazian E, Atyabi F, Ostad SN, Madjd Z, Dinarvand R. Peptide-conjugated liposomes for targeted miR-34a delivery to suppress breast cancer and cancer stem-like population. *J Drug Deliv Sci Technol*. 2020;57:101687.

Wadhwani N, Jatoi I. Overuse of Neo-adjuvant Chemotherapy for Primary Breast Cancer. *Indian J Surg Oncol*. 2020 Mar;11(1):12–4.

Wang H, Cheng G, Du Y, Ye L, Chen W, Zhang L, et al. Hypersensitivity reaction studies of a polyethoxylated castor oil-free, liposome-based alternative paclitaxel formulation. *Mol Med Rep.* 2013;7(3):947–52.

Wang L, Liang T-T. CD59 receptor targeted delivery of miRNA-1284 and cisplatin-loaded liposomes for effective therapeutic efficacy against cervical cancer cells. *AMB Express*. 2020 Mar;10(1):54.

Wang YT, Li B, Li XG, Ma SK, Zhang R, Wu LY. Efficacy and side effect analysis of paclitaxel liposome for neoadjuvant chemotherapy in locally advanced cervical cancer. *Zhonghua Fu Chan Ke Za Zhi*. 2019 Sep;54(9):588–94.

Waterhouse DN, Dragowska WH, Gelmon KA, Mayer LD, Bally MB. Pharmacodynamic behavior of liposomal antisense oligonucleotides targeting Her-2/neu and vascular endothelial growth factor in an ascitic MDA435/LCC6 human breast cancer model. *Cancer Biol Ther.* 2004 Feb;3(2):197–204.

Wu J, Lu Y, Lee A, Pan X, Yang X, Zhao X, et al. Reversal of multidrug resistance by transferrin-conjugated liposomes co-encapsulating doxorubicin and verapamil. *J Pharm Pharm Sci a Publ Can Soc Pharm Sci Soc Can des Sci Pharm.* 2007;10(3):350–7.

Xiong X-B, Huang Y, Lu W-L, Zhang H, Zhang X, Zhang Q. Enhanced intracellular uptake of sterically stabilized liposomal Doxorubicin *in vitro* resulting in improved antitumor activity *in vivo*. *Pharm Res*. 2005 Jun;22(6):933–9.

Xu L, Huang C-C, Huang W, Tang W-H, Rait A, Yin YZ, et al. Systemic tumor-targeted gene delivery by anti-transferrin receptor scFv-immunoliposomes. *Mol Cancer Ther*. 2002 Mar;1(5):337–46.

Xu L, Tang W-H, Huang C-C, Alexander W, Xiang L-M, Pirollo KF, et al. Systemic p53 Gene Therapy of Cancer with Immunolipoplexes Targeted by Anti-Transferrin Receptor scFv. *Mol Med.* 2001;7(10):723–34.

Xu X, Wang L, Xu HQ, Huang XE, Qian YD, Xiang J. Clinical comparison between paclitaxel liposome (Lipusu®) and paclitaxel for treatment of patients with metastatic gastric cancer. *Asian Pacific J Cancer Prev.* 2013;14(4):2591–4.

Yan DH, Chang LS, Hung MC. Repressed expression of the HER-2/c-erbB-2 proto-oncogene by the adenovirus E1a gene products. *Oncogene*. 1991 Feb;6(2):343–5.

Yatvin MB, Weinstein JN, Dennis WH, Blumenthal R. Design of Liposomes for Enhanced Local Release of Drugs by Hyperthermia. *Science* (80-). 1978;202(4374):1290–3.

Yoo GH, Hung MC, Lopez-Berestein G, LaFollette S, Ensley JF, Carey M, et al. Phase I trial of intratumoral liposome E1A gene therapy in patients with recurrent breast and head and neck cancer. *Clin cancer Res an Off J Am Assoc Cancer Res. 2001* May;7(5):1237–45.

Zhang JA, Anyarambhatla G, Ma L, Ugwu S, Xuan T, Sardone T, et al. Development and characterization of a novel Cremophor EL free liposome-based paclitaxel (LEP-ETU) formulation. *Eur J Pharm Biopharm Off J Arbeitsgemeinschaft fur Pharm Verfahrenstechnik* eV. 2005 Jan;59(1):177–87.

Zhao Z, Ukidve A, Kim J, Mitragotri S. Targeting Strategies for Tissue-Specific Drug Delivery. *Cell*. 2020 Apr;181(1):151–67.

Zhigaltsev I V, Belliveau N, Hafez I, Leung AKK, Huft J, Hansen C, et al. Bottom-up design and synthesis of limit size lipid nanoparticle systems with aqueous and triglyceride cores using millisecond microfluidic mixing. *Langmuir*. 2012 Feb;28(7):3633–40.

Zhu L, Chen L. Progress in research on paclitaxel and tumor immunotherapy. *Cell Mol Biol Lett*. 2019;24(1):40.

Zolsketil pegylated liposomal | European Medicines Agency.

Zou Y, Zong G, Ling Y-H, Perez-Soler R. Development of cationic liposome formulations for intratracheal gene therapy of early lung cancer. *Cancer Gene Ther*. 2000;7(5):683–96.

Chapter 5

Liposomes for Gene Delivery: Methods and Application

Neha Jaiswal[1,*]
and Swarnima Pandey[2]
[1]Department of Pharmaceutics, College of Dentistry and Pharmacy,
Buraidah Private College, Buraidah, Kingdom of Saudi Arabia
[2]Department of Pharmaceutics, Apex College of Pharmacy, Rampur, Uttar Pradesh, India

Abstract

Gene-based therapy is a promising field that aims to treat diseases by intentionally modulating gene expression in specific cells. To achieve this modulation, exogenous nucleic acids like DNA, mRNA, small interference RNA (siRNA), microRNA (miRNA), or antisense oligonucleotides are injected. In a relatively short period of time, the active field of gene therapy has advanced quickly into clinical trials. The development of a vector capable of serving as a safe and effective gene delivery vehicle is critical to the success of any gene therapy procedure. The creation of non-viral DNA-mediated gene transfer methods, such as liposomes, has been aided by this. As a result of their huge size and negative charge, these macromolecules are often delivered by carriers or vectors. Liposome-based delivery systems have shown amazing advancements with substantial therapeutic consequences. These include elastic liposomes for topical, oral, and transdermal administration, as well as covalent lipid-drug complexes for improved drug plasma membrane crossing and targeting specific organelles. Long-circulating, stimuli-responsive, nebulised, and stimuli-responsive liposomes were also included. The development of nonviral

* Corresponding Author's Email: neha.jais17@gmail.com.

In: Liposomes
Editors: Usama Ahmad and Anas Islam
ISBN: 979-8-89113-636-6

lipid-based vectors, particularly liposome-based delivery systems, notably liposome-based delivery systems, has advanced gene-based therapy significantly. The advantages of these vectors, such as low immunogenicity, low toxicity, and ease of production, make them promising candidates for clinical applications in gene therapy. Ongoing research in this field will undoubtedly lead to more innovative and effective nonviral lipid-based vectors for therapeutic gene delivery.

Keywords: liposomes, mRNA, DNA, gene therapy, targeted delivery, non-viral vector

1. Introduction

Gene therapy is a new medical modality and technology that allows for precise changes in the human genome. Since genes were identified as the basic building blocks of heredity, medicine has worked to make site-specific modifications in the genome. The scientific community is interested in gene therapy because it has the ability to treat diseases at their root. Gene therapy essentially involves correcting mutated genes or creating specific alterations for therapeutic purposes to improve genetic function. Genetic and bioengineering developments have allowed the use of vectors to deliver extrachromosomal material to particular cells, making this treatment strategy possible. Gene therapy presents a viable path for treating and possibly curing a variety of diseases by utilising these developments. (Lasic 1997; Saffari et al., 2016).

During gene therapy, a functioning gene or other molecule containing genetic information is put into a cell to achieve a therapeutic effect. The gene itself serves as a type of treatment in this situation. The delivery of genetic resources to living cells, such as DNA, RNA, and antisense sequences, to treat genetic abnormalities is the main goal of gene therapy. The need to create secure and effective methods to deliver genes into cells is a barrier to the advancement of gene therapy, even though it has enormous potential as a disease treatment. Because it has benefits over enzyme and protein replacement treatment, gene therapy is often thought of as an alternative. Replacement therapies may face difficulties such as *in vivo* clearance and high manufacturing costs, making gene therapy a viable option for a variety of scarce genetic diseases (Ramamoorth and Narvekar, 2015). The ideal gene delivery technology should be capable of selective targeting, biodegradable, non-toxic, non-inflammatory, non-immunogenic, and stable for storage. Moreover, it should have a high capacity for transporting genetic material, efficient transfection capability, and the ability to be produced on a large scale at a reasonable cost.

Gene therapy entails inserting a functional gene into the genetic material to replace a faulty gene responsible for a certain disease. The key challenge in overcoming the different obstacles in this method is properly delivering the gene into the stem cell. A specialised molecular transporter called a "vector" is used to address this issue. The vector must satisfy a number of requirements, such as accurate targeting, effective release of many genes that are of appropriate sizes for clinical application, evasion of immune system recognition, and the capacity to be purified in significant amounts at high concentrations for widespread manufacturing. After the vector has been incorporated into the patient, it must not cause allergic or inflammatory reactions. It should improve healthy processes, fill in gaps, or stop destructive behaviour. It should also guarantee safety for all involved, including the environment, the healthcare workers handling it, and the patient. In the majority of cases, the vector should be able to express the gene for the duration of the patient's life (Gonçalves and Paiva, 2017; Misra, 2013; Gardlk et al., 2005). Gene therapy vectors currently fall into general classifications, which include viral vectors, nonviral vectors, and engineered vectors. Non-viral vectors encompass chemical, particle, and naked DNA-based systems. These vectors can be administered directly through plasmid DNA or naked DNA, as well as by chemical, physical, or combined methods (Ramamoorth and Narvekar, 2015).

Compared to nonviral techniques, viral vectors show a comparatively high efficacy in transfecting host cells. However, there are a number of disadvantages to using viral vectors. Their immunogenicity, dependency on packaged cell lines, safety issues, possible immunological reactions, lack of cell-specific targeting, and cytotoxicity are some of these. The viral vector (Adenovirus), which caused the first known mortality in a clinical experiment in gene therapy, was blamed for an inflammatory response. Insertional mutagenesis, in which viral DNA is integrated into chromosomes in an abnormal location and disrupts the expression of tumour suppressor genes or activates oncogenes, is another worrying issue associated with viral gene transfer vehicles. This condition results in malignant transformation of cells. Additionally, they are rapidly eliminated from the bloodstream, resulting in limited transfection mainly occurring in "first-pass" organs such as the lungs, liver, and spleen. (Schmidt and Schmidt, 2003; Lollo et al.,

2000). The occurrence of recombination events that result in the creation of a replicating virus is a distant concern. Non-viral vectors offer significant safety benefits compared to viral approaches, as they have demonstrated lower pathogenicity, cost-effectiveness, and straightforward production methods (Ramamoorth and Narvekar, 2015; Ponder, 2000).

The biosafety of using nonviral vectors is the main advantage. However, the use of non-viral gene transfer methods was previously overlooked due to their limited delivery efficiency and resulting low transient expression of transgenes (Gloves and Lipps, 2005). Nonviral vectors, however, have drawn a lot of attention because of their decreased immunotoxicity. Between 2004 and 2013, the use of nonviral vectors has shown a noticeable increase in clinical trials, whereas there was a noticeable decline in the use of viral vectors. An increase in the number of nonviral vector products being tested in clinical trials can be attributed to improvements in efficacy, specificity, duration of gene expression, and safety (Ramamoorth and Narvekar, 2015; Mali, 2013).

Liposomes, which act as efficient transporters, present a promising method for delivering therapeutic materials such as DNA. Liposome-based vectors, unlike viral vectors, can be specifically engineered to safely target specific cells using targeted molecules. Liposomes are a perfect medication delivery device because of their membrane-like biological shape. Additionally, their adaptability enables the insertion of diverse chemicals, boosting their functionality, such as nucleic acids, peptides, antibodies, aptamers, and folic acid. The efficiency of liposomes as carriers is influenced by a variety of elements, including the physicochemical characteristics of their membranes, constituent composition, size, surface charge, and lipid arrangement. Chemical alterations can significantly improve the performance of liposomes in various aspects, such as prolonged circulation in the body, targeted delivery to particular sites, improved penetration into cells, increased contrast for image-guided therapy, and responsiveness to stimuli. Numerous aspects must be considered to successfully deliver a gene using liposomes, and the creation of optimised liposomal systems has the potential to maximise gene transfection, predominantly in *in vivo* applications. Currently, a variety of liposomes, such as charged liposomes, stimuli-responsive liposomes, and PEG-modified liposomes, are employed as vehicles for drug delivery. Each liposome form has unique properties and advantages. (Torchilin, 2005; Allen and Cullis, 2004; Buzzuto and Molinari, 2015).

Due to their exceptional attributes in delivering DNA, liposomes stand out as one of the most potent tools for gene delivery:

1. The immune response is lowered
2. Convenient loading of substantial DNA quantities
3. The stability of DNA in the body is improved.
4. The absence of clearance leads to improved circulation of DNA in the bloodstream.

Liposomal gene therapy could help a variety of diseases, including cystic fibrosis, sickle cell anemia, immune system deficiencies, transmissible viral diseases (HIV, hepatitis), neurological diseases (Parkinson's and Alzheimer's disease) and haemophilia (Balazs and Godbey, 2011; Ibraheem, 2014).

2. Fundamentals Characteristics of Liposomes

2.1. Structure of the Cell Membrane and Liposome

The fluid mosaic concept proposed by Singer and Nicolson (1972) is assumed to be reflected in the structure of the cell membrane, which is mostly made up of lipids and proteins. The phospholipid bilayer, which is where membrane proteins are hidden, is made up of lipid molecules that are only loosely connected to one another by hydrophobic interactions while still retaining some mobility. Bangham et al., discovered in 1965 that phosphatidylcholine from egg yolks created vesicles in water and these vesicles were capable of encapsulating both cations and anions. This vesicle was later given the name liposome. The hydrophilic heads of the phospholipids interact with the surrounding liquid, forming a continuous lipid wall, while the hydrophobic fatty acid tails of the phospholipids associate with each other within the lipid bilayer, preventing water from entering (Figure 1). (Bangham et al., 1965; Singer and Nicolson, 1972).

Liposomes are colloidal vesicular particles made up of amphiphilic molecules that self-assemble. Amphiphiles are compounds that include solubility-varying groups. The hydrophilic portion is frequently referred to as the polar head, whereas the hydrophobic portion is known as the nonpolar tail.

According to their physical characteristics, liposomes are divided into three subgroups: multilamellar vesicles (ML V), large unilamellar vesicles (LUV), and small unilamellar vesicles (SUV). MLVs are distinguished from SUVs and LUVs by their multi-lamellar lipid bilayer, which resembles the bulb of an onion. The MLV sizes ranged from 100 manometers to several micrometres. MLVs are more stable than LUVs and SUVs and are simple to construct. The multilamellar barrier makes it challenging to consistently create ML Vs of the same size and effectively deliver the enclosed transgene to the target cells. LUVs had sizes ranging from 100 nm to 1000 nm. Compared to intravenous gene therapy LUVs are considered inferior because the reticuloendothelial system can easily entrap them, even though they can entrap high-molecular weight molecules. The SUVs had diameters of less than 100 nm. Although small unilamellar vesicles (SUVs) may not be as efficient as multilamellar vesicles (MLVs) or large unilamellar vesicles (LUVs) in encapsulating a large number of larger molecules, their advantage lies in their highly uniform size distribution. The presence of a single lipid bilayer in SUVs makes them more effective in delivering genes to the desired target cells (Maja et al., 2020; Nasairat et al.,2022; Yanagihara et al., 2012).

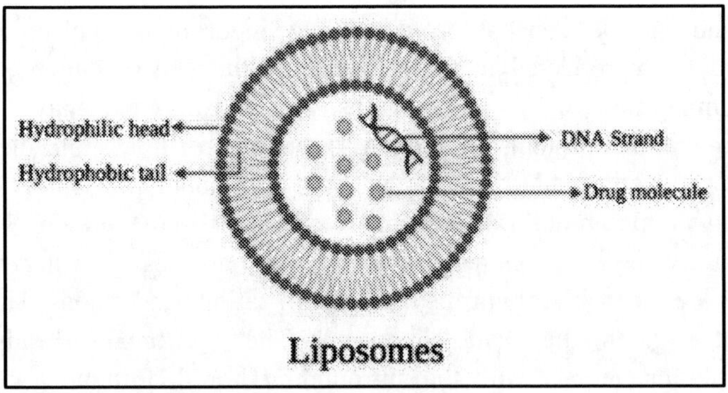

Figure 1. Structure of liposomes.

2.2. Interaction between Liposome Cell Membrane

Studies of the interactions between cell membranes and liposomes have served as models for biological membrane interactions. Four primary forms of interactions between cell membranes and liposomes (Yanagihara et al., 2012).

2.2.1. Inter-Membrane Transfer

Random exchanges of lipids or proteins between liposomes and cell membranes are possible. In contrast to membrane fusion, intermembrane transfer is often exceedingly slow and insufficient for phospholipid transport. However, lipid carrier proteins, which frequently contain lipoproteins, can allow phospholipids to enter the cell membrane more effortlessly. Lipoproteins can also be exchanged with liposomes due to the structural resemblance between liposomes and lipoproteins containing a phospholipid monolayer. The transfer of lipids between liposomes and Golgi membranes, cell plasma membranes, and newly formed liposomes has been demonstrated using fluorescence-labelled lipids, which confirms the transport (Nasairat et al., 2022; Lipsky and Pagano, 1985).

2.2.2. Absorption of Liposomes into Cell Membranes

Some proteins, such as antibodies and glycolipids, can bind to cell membranes. These compounds can interact with particular elements of the cell membrane when they are integrated within the lipid bilayer of the liposome. For example, sialyl Lewisx (LeX), a ligand for L-selectin (a cell adhesion molecule), which is present in liposomes that effectively adhere to teratocarcinoma cells (Eggens et al., 1989). It is crucial to remember that this binding does not always cause the enclosed elements to move into the cell's interior. Additionally, cationic liposomes attach to negatively charged cell membranes via electrostatic interactions.

2.2.3. Endocytosis of Liposomes in the Cell

Following attachment to the cell membrane, the liposomes associated with the coated pits undergo internalisation into coated or uncoated vesicles. Through the endocytosis pathway, liposomes are sorted into secondary lysosomes, where they are exposed to lysosomal enzymes, leading to the degradation of the liposome membrane and encapsulated materials. However, advances in pH-sensitive liposomes have facilitated the release of trapped materials from the endosome compartment into the cytoplasm (Zabner et al., 1995; Khosravi-Darani et al., 2010).

2.2.4. Liposomes and Cell Membrane Fusion

A desirable approach for gene transfer involves the fusion of liposomes with the cell membrane, since this process enables direct internalisation of foreign genetic material into the cell cytoplasm, bypassing the need for lysosomal degradation. Although the occurrence of liposome fusion is widely recognised, fusion between liposomes and cell membranes is relatively rare. For example, it is known that phosphatidylcholine liposomes can fuse with cell membranes, but this fusion process normally takes several days to complete. Due to the ease with which the reticuloendothelial system can trap unattached liposomes, this sluggish fusion rate has considerable negative effects on *in vivo* gene transfer. Therefore, the development of fast fusion mechanisms becomes crucial in the context of liposome-mediated gene therapy.

Many viruses have the ability to join the viral envelope to the cell membrane, which is facilitated by glycoproteins. Fusion of the endosomal and cell membrane is the two main mechanisms of fusion within viral envelopes and cell membranes that have been discovered. Many single-stranded RNA viruses enter cells by receptor-mediated endocytosis, including influenza viruses, alphaviruses, rhabdoviruses, and flaviviruses. When the viral envelope comes into contact with the acidic pH of the endosome, it fuses with the endosomal membrane, allowing the viral DNA to be released into the cell cytoplasm. Viral envelope glycoproteins, on the other hand, allow paramyxoviruses, coronaviruses, and retroviruses to fuse directly with cell membranes (Lamb, 1993). To effectively use gene therapy, the desired genes must be internalised into the cytoplasm without being degraded in the endosome-lysosome route (Yanagihara et al., 2012).

2.3. Toxicity and Antigenicity

Although biodegradable lipids make up the majority of liposomes, liposome toxicity and antigenicity must be constantly considered when developing clinical goods. DNA synthesis in cells is inhibited by stearylamine (SA), phosphatidylserine, and dicetyl phosphate (DCP), while cholesterol, phosphatidylcholine, phosphatidic acid, and phosphatidylglycerol do not. When administered intracerebrally to the mouse brain, DCP and SA have additionally been shown to be harmful lipids. Furthermore, it is understood that while cholesterol hemisuccinate is safe for rats with normal physiologies, it is fatal to hypophysicized rats (Weiner, 1989). Therefore, attention should be paid to clinical liposome trials in patients with abnormalities or disruptions in lipid metabolism. In a recent experiment, the use of a DNA and lipid formulation (dimyristyloxypropyl-3-dimethyl-hydroxyethyl ammonium (DMRIE) and dioleoyl phosphatidylethanolamine (DOPE)) at concentrations 1000 times higher than those used in human gene therapy protocols did not lead to any pathological alterations (San et al., 1993). Hydrophobic proteins that are exposed to exterior membrane of liposomes are often antigenic, but proteins that are entirely trapped are not.

3. Preparation of Liposomes

The liposome formulation is well documented in the literature. Here, we briefly describe some of these issues. Traditionally, there are three ways to create liposomes: mechanical dispersion, solvent dispersion, and detergent solubilisation. These techniques are primarily based on factors such as the composition of the materials utilised for liposome membrane preparation, the solubility of the substance intended for encapsulation (whether it is water-soluble or lipid-soluble), and the size of the resulting liposomes (Bagasra et al., 1999; New, 2017).

3.1. Mechanical Dispersion Method

Various techniques can be employed to create liposomes, including the application of lipid droplets onto the surface of a glass container, the hydratation of lipids by adding an aqueous medium, and the subsequent shaking of hydrated lipids using different methods. These processes lead to the fabrication of multi-lamellar vesicles. The dimension of the liposomes can be adjusted by varying the intensity of shaking. An aqueous solution encapsulated within a lipid membrane often occupies only a small percentage of the total volume necessary for swelling. Consequently, while the output of captured material per millilitre or gm of lipid can be excellent, the mechanical dispersion approach results in wastage of water-soluble molecules intended for entrapment.

3.2. Solvent Dispersion Technique

The technique of liposome preparation, initially described by Batzri and Korn in 1973, involves the dissolution of the lipid components of the liposome membrane in an organic solvent. The lipid solution was then mixed with an aqueous phase containing the desired materials to be trapped within the liposomes. This method can be further classified on the use of water-miscible or water-immiscible organic solvents. The technique yields a significant quantity of unilamellar or multilamellar vesicles of a specific size, and the success of the process relies on the proper mixing of the lipid with the solvent. The degree of encapsulation in the aqueous medium is influenced by the characteristics of the organic solvents employed. One notable benefit of this approach is its simplicity and reduced risk of degradation caused to the sensitive lipids. However, constraints exist as a result of factors such as lipid solubility in organic solvents, the maximum volume of solvent that can be introduced into the system, and the subsequent solvent extraction. These factors collectively contribute to restricted dispersion of lipids, potentially resulting in a low percentage of encapsulation (Batzri and Korn, 1973).

3.3. Detergent Solubilisation Technique

This method relies on the dissolution of lipids in a detergent, which is subsequently combined with the aqueous phase that contains the intended material for encapsulation. This procedure initiates the creation of micelles, a pivotal stage in the liposome preparation process using this particular approach. The nature (ionic, nonionic, or amphoteric) and concentration of the detergent and lipids used define the size, form, and composition of the micelles (bilayer, lamellar, or spherical). The solubilisation approach of detergents is ideal for producing liposomes containing lipophilic proteins incorporated into the lipid membrane. However, it is important to note that detergent methods generally exhibit inefficiency in terms of the percentage of material successfully entrapped within liposomes (Pradhan et al., 2015).

4. Conventional Liposome Systems

4.1. Simple Liposomes

Cholesterol and phosphatidylserine have been combined to form negatively charged liposomes, which have been utilised for gene delivery *in vitro*. The thymidine kinase (TK) gene, which lacks TK activity, was inserted into mouse L cells for this investigation. The TK gene was shown to be transiently expressed in approximately 10% of total L cells. Fourteen days after transfection, approximately 0.02% of total L cells were identified as stable transformants. This simple liposome-based method demonstrated a three-fold higher efficiency for transient expression compared to the conventional calcium phosphate gene transfer method. However, it showed approximately half the effectiveness of stable expression in comparison to the calcium phosphate method (Schaefer-Ridder et al., 1982; Yanagihara, 2012).

4.2. Cationic Liposomes

Cationic liposomes are the nonviral vectors that have received the most research attention. Currently, preclinical and clinical trials are being conducted to investigate its potential to deliver therapeutic genes. This is accomplished using cationic lipid-pDNA complexes, also known as cationic lipoplexes. Cationic lipoplexes can carry huge amounts of polynucleotides into somatic cells, are simple and inexpensive, and are composed of nontoxic and non-immunogenic precursors. These substances can also be easily changed within the laboratory to include novel biological functionalities or generate new combinations for the purpose of testing their *in vivo* gene transfer capabilities. (Ropert, 1999; Schmidt and Schmidt, 2003)

Due to the plasmid size and inadequate transfection technique, it can technically be difficult to encapsulate DNA in typical liposomes. Based on this, cationic lipids and PE-based alternative technologies emerged in late 1980s. The goal was to trap plasmids more effectively and transfer DNA to cells by neutralising negative charge plasmids with positive charged lipids. In general, this is a straightforward process that involves combining DNA and cationic lipids before adding them to cells. As a result, aggregates made of DNA and cationic lipids are created. (Felenger et al., 1987; Ropert, 1999).

The synthesis of the cationic lipid DOTMA (N-(1-(2,3-dioleyloxy) propyl)N, N,N-trimethylammonium chloride) was carried out as explained by Felgner et al., in 1987. This lipid can spontaneously form multilamellar vesicles (MLV) on its own or in combination with other neutral lipids. These MLVs can then be sonicated to produce small unilamellar vesicles (SUVs). When DNA is introduced, it readily interacts with DOTMA, leading to the development of DNA complexes in which 100% of the DNA becomes associated. Ionic interactions between the positively charged head group of DOTMA and the negatively charged phosphate groups of DNA are thought to cause complex formation. DOTMA is commercially available (sold as Lipofectin by Gibco-BRL, Gaithersburg, MD) and is often combined with dioleyl phosphatidyl ethanolamine (DOPE). It is commonly used for transfection of many cell types (Brigham et al., 1989; Innes et al., 1990; Brant et al., 1991; Li et al., 1992). In an aqueous environment, DOTMA can form lipid bilayers, either by itself or in conjunction with other phospholipids. In addition, DOTMA easily establishes complexes with nucleic acids such as DNA and RNA by simply mixing. The liposome-to-DNA ratio significantly influences the efficiency of gene expression. Although liposomes are electrostatically drawn to the cell membrane, it is crucial for liposome-DNA complexes to maintain a positive charge overall to ensure effective binding to negatively charged cell membranes and prevent their exclusion due to an excess of negatively charged DNA. The Lipofectin® system has been proven to be highly valuable for gene transfection due to its user-friendly nature, broad range of applicable host cells *in vitro*, and the ability to accommodate large DNA molecules.

According to Felgner and Ringold's hypothesis (1989), four cationic liposomes envelope linear DNA (Felhner and Ringold, 1989). On the other hand, a plasmid has the ability to take on a supercoil structure, which reduces its size by 41% in comparison to the overall DNA length. This structure consists of a central core branch with four additional branches that stem from it. Consequently, it is believed that supercoiled DNA-liposome complexes are more condensed than the complexes of a linear DNA-liposome. In addition, these condensed complexes have the potential to aggregate with each other (Smith et al., 1993).

In a fluorescence-based study of DNA-liposomes, Lipofectin demonstrated efficient fusion with target cell membranes, resulting in over 99% of the cells containing the plasmid. According to research on plasmids created in the nucleus of each cell, less than 1% of them are intact (Felhner and Ringold, 1989). The technology of lipofection gene transfer appears to offer some advantages for directly expressing foreign genes *in vivo*, particularly in the arteries and lungs. Cationic liposome-mediated gene therapy has been used in animal models of cystic fibrosis by instillation and nebulization (Hyde et al., 1993; Alton et al., 1993). After the human cystic fibrosis transmembrane regulator gene was transferred using both methods, the ion transport deficiency was partially corrected.

In clinical trials, liposomes have been examined as potential therapies for treating malignant melanoma. In this study, HLA-B7, a protein belonging to the major histocompatibility complex (MHC), was administered to five patients with stage IV melanoma who lacked HLA-B7. The therapy aims to stimulate an immune response by introducing a foreign antigen, triggering cytotoxic T cells to target the tumour using a foreign MHC (class I) signal. A patient experienced tumour regression (Nabel et al., *1993*). A variety of metabolisable quaternary ammonium salts have been developed to reduce the cytotoxic impact of DOTMA. These salts exhibited an efficacy comparable to that of lipofectin when combined with DOPE (Leventis and Silvius, 1990; Ropert, 1999). However, an additional disadvantage is that the lipofectin system is hindered by negatively charged serum components, such as sulfated proteoglycans.

An essential aspect of transfection is the compaction of DNA to enhance its cellular penetration. Cationic amphiphiles capable of compacting genomic DNA, specifically lipopolyamines, were investigated. Transfectam™ (DOSG) has proven to be extremely effective in transfecting a wide range of animal cells (Behr et al., 1989; Barthel et al., 1993; Staedel et al., 1994). These amphiphiles were found to be capable of successfully condensing DNA into stable particles. Various detergents with different structures, such as cetyltrimethylammonium bromide (CTAB) and dodecyltrimethylammonium bromide (DDTAB), have been compared in conjunction with DOPE. Among them, DDTAB showed the highest promising results, leading to the patenting of the DDTAB/DOPE formulation (TransfectACETM). According to Farhood et al., DOPE plays a crucial role in cationic liposome-mediated gene transfer and has been widely employed. The mechanism of DNA/cationic lipid by cells is believed to be associated with endocytosis, and DOPE is proposed to aid in the release of DNA into the cytosol, particularly in formulations sensitive to pH (Farhood et al., 1995).

According to the Cryo-TEM study, an overabundance of lipids can trap DNA molecules between the lamellae, leading to the formation of clusters and aggregated multilamellar structures. The morphology of DNA-lipid complexes does not appear to be affected by the use of surfactant. The

DOPE-containing system also produced more condensed aggregates when compared to formulations using egg lecithin. In their model, Templeton et al., used cholesterol, N-1(2,3-dioleyloxy)propyl, N, N, N-trimethylammonium methyl sulphate (DOTAP) and DOTAP to explain how DNA-lipid complexes are put together. According to their model, DNA forms closed structures that protect DNA by adhering to the invaginated and tubular liposomes through electrostatic interactions. (Templeton et al., 1997).

According to Farhood et al., endocytosis is the main mechanism through which cells take complexes of DNA-lipids during transfection. Through endocytosis, the complex attached to the cell surface is internalised into the endosomes and lysosomes. A sizable fraction of DNA is expected to degrade within these compartments (Farhood et al., 1995). Hui and Zhao supported this idea by suggesting that endocytosis, rather than direct fusion of the complex with the plasma membrane, is the most common method for the entry of DNA into CHO cells (Hui and Zao, 1995).

4.2.1. Delivery of the Cationic Nanolipoplex Gene with Divalent Cations

Cationic lipoplexes, a type of complex comprising cationic liposomes and DNA molecules, are frequently employed as transfection vectors in both *in vitro* and *in-vivo* settings (Alton and Geddes, 1995; Liu et al., 1995; Smith et al., 1993). The makeup of target cell and the physical properties of the liposomes (membrane fluidity, charge, and size) affect how the liposomes interact with the membrane (Khosravi-Darani et al., 2010).

Endocytosis is the primary process by which liposome-DNA complexes interact with cells. Liposome stability and cell interaction potential are affected by serum (Bonté and Juliano, 1986). Additionally, the transfection effectiveness of cationic lipoplexes was adversely affected by the presence of serum (Oku et al., 2001). Positively charged liposomes exhibit a higher affinity for serum proteins than negatively charged ones, according to studies (Deol and Khuller, 1997). There are a number of reasons why cationic lipoplex transfection is not very successful. These include decreased entry into the nucleus, concerns about toxicity, inadequate lipoplex escape from endosomes during endocytosis, DNA destruction by intracellular nucleases, and increased leakage caused by serum proteins (Lechardeur et al., 1999; Filion and Philips, 1997). Several techniques have been created to enhance the transfection efficacy of cationic liposomes. These include the utilisation of targeting ligands, such as nuclear localisation signals, helper lipids, fusogenic peptides, and DNA-condensing agents. With these initiatives, the difficulties related to cationic lipoplex transfection can be resolved, and overall effectiveness will increase (Gao and Huang, 1996; Hagstrom et al., 1996; Ibáez et al., 1996; Mizuguchi et al., 1996; Li and Huang, 1997). Calcium phosphate was also used to boost the efficacy of cationic lipoplexes during *in vitro* transfection by increasing endocytosis and its entry into the nucleus. The increased transfer of genes to different cell lines through calcium and phosphate is the reason for better transfection efficiency. These results were exclusive to calcium ion transfection and did not apply to transfection with other divalent metal ions. (Khosravi-Darani et al., 2010; Lam and Cullis, 2000).

4.2.2. Transfer of Noncationic Nanolipoplex Gene with Divalent Cations

The main limitation as a therapeutic tool lies in their toxic nature. Cationic amphiphiles or detergents, of which they are composed, exhibit cytotoxic properties that induce cell death in both *in vitro* and *in vivo* environments. Furthermore, these liposomes are inactivated in serum (Kharakoz et al., 1999; Lappalainen et al., 1994). The causes of toxicity of cationic liposomes have not yet been fully determined. According to some theories, cytotoxicity is caused by the mixing of cationic and anionic lipids in the membranes of cell organelles such as mitochondria. Another proposed mechanism of cationic lipid-mediated toxicity in the lungs is the involvement of reactive oxygen intermediates.

The focus of researchers has been on the manufacturing of new cationic lipids and developing better formulations that can enhance the efficacy of cationic liposomes while reducing their toxicity levels, ideally eliminating them altogether (Gao and Huang, 1996; Anwer et al., 2000). Using zwitterionic or anionic lipid vesicles as gene transfer vectors is an alternative to cationic liposomes. These species have longer circulatory durations and a variety of clearance characteristics, making them substantially less dangerous to target cells. (Gulati et al., 1998; Tardi et al., 1996). Unfavourable electrostatic interactions arise between anionic or zwitterionic lipids and negatively charged DNA (Perrie and Greoriadis, 2000; Patil and Rhodes, 2000) owing to repulsive electrostatic forces. Consequently, these interactions lead to low entrapment of DNA molecules (lakkaraju et al., 2001) and poor transfection efficiency. To address the issue of low DNA entrapment within anionic or zwitterionic liposomes, calcium addition was found to be effective in neutralising the negative charge of the DNA phosphate groups. A method that utilises divalent cations to facilitate the incorporation of DNA molecules into anionic liposomes has been reported. Morphological characterisation was conducted to examine the structure of ternary complexes comprising liposome-Ca^{2+}-DNA (Zareie et al., 1997; Mozafari et al., 1998).

Furthermore, a lot of research has been done to understand how calcium interacts with DNA in liposomes that include the phosphate head group of the lipid and the phosphate group of DNA; calcium is thought to both zwitterionic and anionic lipids (Mozafari and Hasirci, 1998). Calcium is believed to act as a bridge between the phosphate head group of the lipid and the phosphate group of DNA. According to a different hypothesis, calcium interacts with the phosphate head group in a 1:2 ratio (Ca2+:phosphatidylcholine), causing a positively charged amine to form in the phosphatidylcholine head group. This positively charged amine then connects with negatively charged DNA (MacManus et al., 2003a; MacManus et al., 2003b). In addition to helping DNA molecules condense, calcium ions also form complexes with nitrogen and oxygen atoms located at positions 7 and 6, respectively, of the guanine molecule. Utilisation of calcium ions offers additional advantages, such as improved entrapment of DNA within liposomes through calcium-mediated binding to DNA liposomes. Provides a broader range of calcium concentrations for DNA condensation while minimising cell toxicity. Furthermore, it facilitates the effortless release of free DNA from Ca2+-DNA complexes when exposed to high sodium concentrations.

4.3. pH-Sensitive Liposome Strategy

Diverse liposome compositions have demonstrated potent affinity for cell surfaces. Dioleylphosphatidylethanolamine (DOPE) has been reported to be the most efficient lipid for *in vitro* gene transfection of these liposomes, particularly in pH-sensitive liposomes or as a lipid helper in cationic liposomes (Wang and Huang, 1987; Felgner et al., 1987). The membrane fusion promoter is believed to involve Phosphatidylethanolamine (PE), a lipid that undergoes changes when subjected to acidification. (Litzinger and Huang, 1992). In addition, cholesterol-containing foods are crucial for maintaining the stability of these liposomes. The interactions of liposomes with cells may be significantly influenced by their chemical composition. For effective cell capture, the size and cell type are critical factors. In general, different endocytosis systems absorb liposomes. Professional phagocytes, such as neutrophils and macrophages, can actively phagocytose liposomes of varying sizes and charges.

4.3.1. The Vesicular Pathway for Cellular Uptake

Liposomes are internalised into endosomes after attachment to the cell surface, where they come into contact with a pH that is more acidic than that of the surrounding environment. In general, the internal pH of the early endosomes was 6.50 (Mellman et al., 1986; Schmid et al., 1988). Through maturation or vesicular fusion, their contents are transported into a more acidic environment. The final endosome environment, with an internal pH of 5.5 to 6.0, was reached ten to fifteen minutes after absorption. Acidification of the lysosome, which is the ultimate endocytic compartment, takes a minimum of 20 minutes after uptake to reach a pH of 5.0 or below. The lysosome is the primary compartment that degrades via the endocytic route. Conventional pH-insensitive liposomes are transported to lysosomes, where they are broken down. Once inside the cell, it is crucial that plasmid liposomes prevent aggregation within specific cellular compartments, such as lysosomes. To prevent this disintegration, pH-sensitive liposomes have been proposed (Wang and Huang, 1987). pH-sensitive liposomes were developed based on the hypothesis that viruses fuse with the endosomal membrane through a protein at pH 5-6, delivering their genetic material to the cytosol before reaching lysosomes (Chu et al., 1990; Connor and Huang, 1986).

Phosphatidylethanolamine (PE) lipids are frequently utilised in the creation of pH-sensitive liposomes. These lipids belong to a group that tends to form non-bilayer structures, such as the inverted hexagonal phase, when isolated. Several stabilizers with titratable acid headgroups, including oleic acid (OA), palmitoyl homocysteine (PHC), and cholesterol hemisuccinate (CHEMS), were used to stabilize PE in the lamellar phase of liposomes. Liposomes made of DOPE / OA / chol can introduce an exogenous TK gene into mouse Ltk cells, which lack thymidine kinase (TK). In this study, pH-sensitive liposomes exhibited eight-fold higher gene delivery efficiency compared to pH-insensitive liposomes. On the contrary, minimal transfection was observed when free plasmid DNA.

Zhou et al., prepared liposomes made of dioleylphosphatidylcholine / DOSG (pH insensitive formulation) or DOPE/dioleylsuccinylglycerol (DOSG) using pUCSV2 CAT DNA. This demonstrates a connection between transfection activity and acid sensitivity. The most effective way to transfect cells was with pH-sensitive DOPE/DOSG liposomes. (Zhou et al., 1992).

Legendre and Szoka investigated the efficiency of transfection mediated by pH-sensitive, pH-insensitive, and cationic (DOPE / doleyloxypropyl-trimethylammonium bromide (DOTMA) liposomes using two different genes and five different cell lines. Cationic liposomes resulted in the highest levels of transfection in all tested cell types. The gene transfer efficiency of pH-sensitive liposomes was found to be 1-30% of that achieved with DOPE / DOTMA, while pH-insensitive liposomes did not induce transfection (Legendre and Szoka et al., 1992).

Ropert et al., conducted a study in which they incorporated antisense oligonucleotides into pH-sensitive liposomes to specifically target the env gene of the murine Friend retrovirus, aiming to inhibit its proliferation. They proposed that the improved activity of the oligonucleotides within the pH-sensitive liposomes was not solely due to destabilisation of the DOPE liposome bilayer, but rather to increased interaction between the pH-sensitive liposomes and the target cells. The researchers found that the viral inhibition efficacy achieved with oligonucleotides encapsulated in pH-sensitive liposomes was only twice as effective as that achieved with oligonucleotides encapsulated in nonpH-sensitive liposomes (Ropert et al., 1992; Ropert et al., 1993). A comparison between pH-sensitive and pH-insensitive liposomes revealed that the former exhibited a twofold increase in cellular uptake and faster absorption by cells, which can lead to improved activity (Chu et al., 1990).

4.4. Hemagglutinating Virus of Japan (HVJ or Sendai Virus)-Liposomes

4.4.1. Principles

Liposome-mediated gene transfer presents two major challenges. The first challenge involves direct delivery of DNA into the cytoplasm to prevent its degradation. To address this problem, Kaneda et al., utilised the hemagglutinating virus of Japan (HVJ) to facilitate the fusion of liposomes with cell membranes, thereby allowing direct delivery of DNA into the cell cytoplasm (Kaneda et al., 1987). The second challenge pertains to the efficient delivery of DNA to the nucleus. To circumvent this barrier, the cointroduction of DNA with a protein known as high mobility group-1

(HMG-1), which promotes DNA migration into the nucleus and increases gene expression, was investigated (Kaneda et al., 1995).

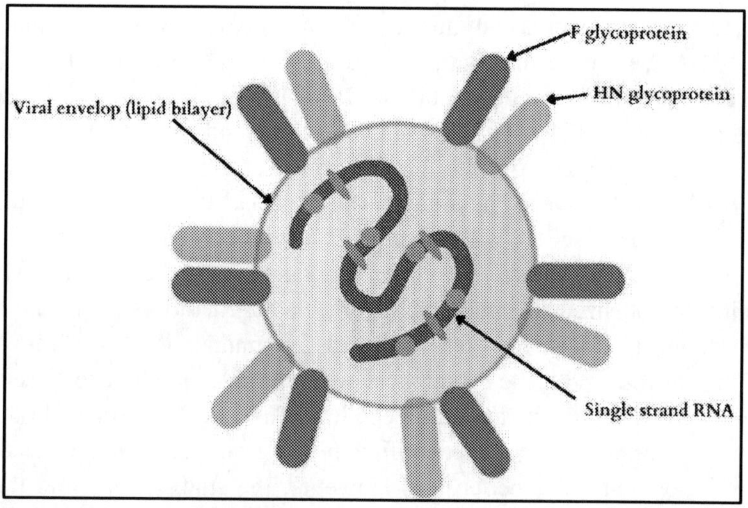

Figure 2. Structure of HVJ-liposomes.

HVJ belongs to the paramyxovirus family, characterised by the presence of two glycoproteins, namely the HN glycoprotein and the F glycoprotein (Figure 2), in its viral envelope. The HN glycoprotein possesses neuraminidase activity that helps identify glycoproteins and glycolipids containing sialic acid, thus facilitating receptor binding on the cell surface. The F (fusion) glycoprotein, on the other hand, exhibits fusion activity in both neutral and acidic pH environments. Although the virus-to-cell fusion mechanism of is still unknown, it is thought that when the HN glycoprotein is recognised and bound to the cell membrane, the F glycoprotein goes through conformational changes that cause hydrophobic fusion peptides to be released and enter the target cell membrane (Lamb, 1993). In terms of safety for future clinical applications, it is important to note that HVJ does not cause disease in humans and can be completely deactivated through appropriate ultraviolet irradiation while still retaining its fusion capability.

4.4.2. In Vivo Transfection by HVJ-Liposome

Numerous examples of gene transfection using HVJ liposomes have been observed both *in-vitro* and *in vivo.* The liver was the main target organ in the early experiments. To provide an example, the human insulin gene was co-delivered into the liver of adult rats via HVJ, using red blood cell ghosts in place of liposomes, and accompanied by HMG-1 or bovine serum albumin (BSA). Compared to BSA complexes, the presence of HMG-1 protein caused a ten-fold increase in human insulin gene. Following injection, DNA and RNA derived from human insulin gene, as well as the resulting expressed insulin, were detectable for 10 to 14 days before quickly depleted (Kaneda *et at.,*1989b). The HVJ-liposome method was used to inject the hepatitis B virus (HBsAg) into the adult rat liver, resulting in the expression of HBsAg in the liver. The antigen was detected in serum for a period of nine days after injection, with the highest level recorded on day two (Kato et al., 1991a). Furthermore, rat serum contains antibodies against HBsAg, leading to localized necrosis and liver cell degeneration with lymphocyte infiltration. This system offers a model for animal hepatitis (Kato et al., 1991b). In the study, the researchers used the HVJ method to observe the manifestation of the human renin gene in the livers of adult rats. On the fifth day, active human renin was identified in the plasma of these rats, leading to the development of hypertension. However, the study also found that this hypertension could be treated with a specific human renin inhibitor (Tomita et al., 1993). The HVJ-liposome technique has successfully facilitated the transfer of numerous genes to different animals *in vivo* without observed damage to organs. Furthermore, the initial data suggested that repeated injections in animals are safe. As a result, although the expression achieved through HVJ liposomes is temporary, it is relatively safe because it does not involve integration into chromosomes.

4.4.3. Antisense Oligonucleotide Delivery

Studies using the HVJ liposome have shown impressive results in the transfer of antisense oligonucleotides (ODNs). Researchers discovered that FITC-labelled ODNs were concentrated in the cell nucleus five minutes after transfer and could be seen for more than three days using HV)-liposomes. Furthermore, the HVJ liposome approach was 50 times more effective than Lipofectin in reducing DNA synthesis in vascular smooth muscle cells (VSMC) through antisense ODNs that block basic fibroblast growth factor. Recently, proliferating cell nuclear antigen (PCNA) and cdc3 kinase antisense ODNs have been studied. *In vitro*, cotransfection of antisense cdc 2 kinase ODNs and antisense PCNA ODNs effectively suppressed VSMC growth. However, individually, antisense cdc 2 kinase ODN and antisense PCNA ODNs had no impact on VSMC growth. *In vivo*, when antisense PCNA and cdc 2 kinase ODN were co-transfected into arteries damaged by angioplasty, a significant decrease in cdc2 and PCNA mRNA was observed, and neointima formation was completely prevented for a duration of up to 8 weeks (Morishita et al., 1993).

The injury caused by balloon angioplasty is known to cause an elevation in the enzyme cyclin-dependent kinase 2 (cdk 2) that regulates the cell cycle. To counteract this, HVJ liposomes were used to inject antisense cdk 2 ODNs into the damaged artery. With inhibition, cdk 2 mRNA decreased markedly (by 60%) during neointima development. In the end, the combined transfection of antisense cdk 2 and antisense cdc 2 ODN led to the complete inhibition of the development of neointima. However, using antisense cdc 2 ODNs alone resulted in inhibition of neointima development by up to 40% (Morishita et al., 1994).

4.5. Liposome-Mediated Nucleic Acid Transfer

In both *in vitro* and *in vivo* studies, liposomes are commonly utilised. In particular, cationic liposomes have shown efficient gene delivery *in vitro* and stimulated gene expression *in vivo*, making them suitable for implementation in gene therapy (Canonico et al., 1994; Yoshimura et al., 1992; Zhu et al., 1993). Cationic liposome DNA complexes are used for the transportation of the cystic fibrosis gene, specifically the CFTR gene (cystic fibrosis transmembrane conductance regulator), which is accountable for this condition. (Hyde et al., 1993). Gene therapy studies have shown the feasibility of using cationic lipid-mediated gene transfer to treat cystic fibrosis in humans (Canonico et al., 1994; Logan et al., 1995). To protect the lungs of rabbits from endotoxin-induced damage and hypertension, a combination of cationic liposomes and a recombinant prostaglandin GIH (PGH) gene complex was used to introduce genes into the lung (Conary et al., 1994).

A cellular immune response was observed along with gene transport and expression upon injection of liposome-DNA formulations directly into tumour masses (Plautz et al., 1993). The use of liposomes for the delivery of the human leukocyte antigen gene, HLA B7, has been reported to result in the cure of melanoma (Hofland and Huang, 1995). The degree of inhibition of cancer cell proliferation in gene therapy seems to depend on the amount of DNA complex delivered to cells, as well as the ratio of lipid to DNA (Kulkarni et al., 2021).

Cationic lipid-DNA complexes delivered through catheters have shown *in vivo* transduction and expression in arterial walls and endothelial cells, making them a useful tool for targeting various organs, including the gastrointestinal system and endothelial tissue, through liposome-mediated gene transfer (Nabel et al., 1993; Bagasra et al., 1999). The ability to transfer genes to diverse cells of the gastrointestinal tract using a liposome-lacZ marker gene opens up the possibility of studying gastrointestinal physiology and developing gene therapies for these diseases (Scmid et al., 1994). Liposomes can also be utilised to transfer genes to the liver, according to some research (Leibiger et al., 1991). Cationic liposomes have been demonstrated to be more efficient than other traditional methods to transfect primary cultured murine hepatocytes (Watanabe et al., 1994).

Liposomes can transport antisense oligonucleotides and ribozymes inside cells to alter viral or cellular gene expression. Several obstacles confront antisense technology, including oligonucleotide breakdown by nucleases and limited membrane permeability. The availability of antisense nucleotides within a cell, required to achieve antisense action, is influenced by various factors. Studies have indicated that intracellular degradation of antisense oligonucleotides can be prevented by using liposomal carriers, which allows their successful transportation into the cell (Rinaldi and Wood, 2018; Dean and Bennett, 2003). Oligonucleotides encapsulated in antibody-targeted liposomes (immunoliposomes) have been shown to be particularly effective in suppressing HIV-1 replication in chronically infected CEM cells (Nsairat et al., 2023). Similarly, within ribozyme technology, the utilisation of liposome-ribozyme complexes presents an alternative approach to overcome the issue of intracellular degradation. The delivery of anti-HIV-1 drugs to lymphoid tissues, which are significant sites of viral production, was successfully achieved using liposome-encapsulated 2,' 3'-dideoxyinosine, an enzyme inhibitor that targets reverse transcriptase (Lu et al., 1994; Filipczak et al., 2020). Due to the potential risks of systemic localisation of delivered DNA, the utilization of processed mRNA in gene therapy techniques is also being evaluated. The one possible choice for transferring processed mRNA is liposomes (Robinson et al., 2018).

5. Prospects for Liposome-Mediated Gene Therapy

Considerable effort is currently focused on the development of techniques for targeting specific organs, enhanced cytoplasmic uptake of encapsulated genes, successful nuclear gene targeting, and long-term gene expression in liposome-mediated *in vivo* gene therapy. There are two main types of gene therapy: compensatory therapy, which seeks to replace enzymes or proteins that are lacking, and preventive therapy, which involves the inhibition of conditions such as cancer or hypertension. Compensatory therapies typically require high levels of systemic gene expression, whereas preventive therapies require more precise treatments that are specific to particular organs or sites to inhibit the expression of unregulated genes. The HVJ-liposome method enables organ targeting by altering the administration route, although greater precision in organ targeting can be achieved by incorporating antibodies, receptor-binding proteins, or cell adhesion recognition molecules into liposome lipid bilayers. In addition, immunosomes and virosomes can be created by embedding viral proteins in the lipid bilayer.

The HVJ fusion activity was utilised to improve gene transfer into the cytoplasm. To minimise the antigenicity of viral protein, it may be crucial to incorporate only the virus fusion protein into the lipid bilayer for liposome-cell fusion. Although transfected animals do not show any evidence of circulating antibodies against HVJ, this approach may be preferable. It will be crucial to employ recombinant technologies to purify, characterise, and synthesize viral fusion protein systems, given their importance. HMG-l has effectively facilitated the translocation of DNA complexes from the cytoplasm to the nucleus, although its mechanism is not yet fully grasped. Understanding the uptake mechanism and the role of stimulated gene transcription will be critical. Creating novel gene expression systems will require more efficient nuclear proteins and/or more selective transcription factors as significant tools.

There are three potential approaches to achieve stable, regulated, and replicable transgenes for long-term gene expression. First, since liposomes can encapsulate large nucleic acids and complex proteins, it may be feasible to introduce (artificial) chromosome-like structures into the nucleus. These structures would contain promoters, enhancers, and transcriptional factors, are safeguarded from nucleases, and ideally possess elements that enable the replication of the transgene in sync with the host cell's proliferation cycle. Second, targeted replacement of damaged gene fragments could facilitate natural regulation of the corrected gene. To accomplish this, the liposome gene transfer method must incorporate *in vivo* site-specific integration or homologous recombination activities *in vivo*. Third, it is known that some viral genomes replicate in the cytoplasm. Because excessive proliferation causes in host cell death, this replication has restrictions in the host cell. Liposomes could become effective carriers for cytoplasmic gene therapy if we can characterise and harness the cytoplasmic replication mechanism. In summary, by resolving and adapting certain intricate molecular processes, liposomes have the potential to evolve into optimal vector systems for gene therapy.

References

Allen, T. M., Cullis, PR. Drug delivery systems: entering the mainstream. *Science*. 2004; 303(5665):1818-22.

Alton, EW, Geddes, DM. Gene therapy for cystic fibrosis: a clinical perspective. *Gene therapy*. 1995; 2(2):88-95.

Alton EW, Middleton PG, Caplen NJ, Smith SN, Steel DM, Munkonge FM, Jeffery PK, Geddes DM, Hart SL, Williamson R, Fasold KI. Noninvasive liposome-mediated gene delivery can correct the ion transport defect in cystic fibrosis mutant mice. *Nature genetics*. 1993; 5(2):135-42.

Anwer K, Bailey A, Sullivan SM. Targeted gene delivery: a two-pronged approach. *Crit Rev Ther Drug Carrier Syst* 2000; 17(4):377-424.

Bagasra, O., Amjad, M., Mukhtar, M. Liposomes in gene therapy. *Gene Therapy: Principles and Applications*. 1999:61-71.

Balazs, D. A., Godbey, WT. Liposomes for use in gene delivery. *Journal of drug delivery*. 2011; 2011.

Bangham, AD, Standish, MM and Watkins, JC. Diffusion of univalent ions across the lamellae of swollen phospholipids. *J Mol Biol*. 1965; 13:238-52.

Barthel F, Remy JS, Loeffler JP, Behr JP. Optimisation of gene transfer with lipospermine-coated DNA. *DNA and cell biology*. 1993; 12(6):553-60.

Batzri S, Korn ED. Single bilayer liposomes prepared without sonication. *Biochimica et Biophysica Acta (BBA)-Biomembranes*. 1973; 298(4):1015-9.

Behr JP, Demeneix B, Loeffler JP, Perez-Mutul J. Efficient gene transfer into mammalian primary endocrine cells with lipopolyamine-coated DNA. *Proceedings of the National Academy of Sciences*. 1989; 86(18):6982-6.

Bonté F, Juliano RL. Liposome interactions with serum proteins. *Chemistry and physics of lipids*. 1986; 40(2-4):359-72.

Bozzuto, G. and Molinari, A. Liposomes as nanomedical devices. *International Journal of nanomedicine*. 2015; 10:975.

Brant M, Nachmansson N, Norrman K, Regnell I, Bredberg A. Propagation of shuttle vector plasmid in human peripheral blood lymphocytes facilitated by liposome-mediated transfection. *DNA and Cell Biology*. 1991; 10(1):75-9.

Brigham KL, Meyrick B, Christman B, Berry Jr LC, King G. Expression of a prokaryotic gene in cultured lung endothelial cells after lipofection with a plasmid vector. *Am J Respir Cell Mol Biol*. 1989;1(2):95-100.

Canonico AE, Conary JT, Meyrick BO, Brigham KL. Aerosol and intravenous transfection of the human alpha 1-antitrypsin gene into the lungs of rabbits. *American Journal of respiratory cell and molecular biology*. 1994;10(1):24-9.

Chu CJ, Dijkstra J, Lai MZ, Hong K, Szoka FC. Efficiency of cytoplasmic delivery by pH-sensitive liposomes to cells in culture. *Pharmaceutical research*. 1990;7:824-34.

Conary JT, Parker RE, Christman BW, Faulks RD, King GA, Meyrick BO, Brigham KL. Protection of rabbit lung from endotoxin injury by *in vivo* hyperexpression of the prostaglandin G/H synthase gene *in vivo*. *The Journal of Clinical Investigation*. 1994;93(4):1834-40.

Connor J, Huang L. pH-sensitive immunoliposomes as an efficient and target-specific carrier for antitumor drugs. *Cancer research*. 1986;46(7):3431-5.

Dean, NM, Bennett, CF. Therapeutics based on antisense oligonucleotides for cancer. *Oncogene*. 2003;22(56):9087-96.

Deol P, Khuller GK. Lung-specific stealth liposomes: stability, biodistribution, and toxicity of liposomal antitubercular drugs in mice. *Biochimica et Biophysica Acta (BBA)-General Subjects*. 1997;1334(2-3):161-72.

Eggens I, Fenderson B, Toyokuni T, Dean B, Stroud M, Hakomori SI. Specific interaction between Lex and Lex determinants: a possible basis for cell recognition in preimplantation embryos and in embryonal carcinoma cells. *Journal of Biological Chemistry*. 1989;264(16):9476-84.

Farhood H, Serbina N, Huang L. Role of dioleoyl phosphatidylethanolamine in cationic liposome-mediated gene transfer. *Biochimica et Biophysica Acta (BBA)-Biomembranes*. 1995;1235(2):289-95.

Felgner, P. L., Gadek, TR, Holm, M., Roman, R, Chan, HW, Wenz, M., Northrop, J. P., Ringold, GM, Danielsen, M. Lipofection: a highly efficient, lipid-mediated DNA-transfection procedure. *Proceedings of the National Academy of Sciences*. 1987;84(21):7413-7.

Felgner PL, Ringold GM. Cationic liposome-mediated transfection. *Nature*. 1989;337(6205):387-388.

Filion MC, Phillips NC. Toxicity and immunomodulatory activity of liposomal vectors formulated with cationic lipids toward immune effector cells. *Biochimica et Biophysica Acta (BBA)-Biomembranes*. 1997;1329(2):345-56.

Filipczak N, Pan J, Yalamarty SS, Torchilin VP. Recent advances in liposome technology. *Advanced drug delivery reviews*. 2020;156:4-22.

Gao X, Huang L. Potentiation of cationic liposome-mediated gene delivery by polycations. *Biochemistry* 1996;35(3):1027- 36.

Gardlk R, Pálffy R, Hodosy J, Lukács J, Turna J, Celec P. Vectors and delivery systems in gene therapy. *Med Sci Monit*. 2005;11(4):RA110-21.

Gloves DJ, Lipps HJ. Towards safe non-viral therapeutic gene expression in humans. *Nat Rev Genetics*. 2005;6:299-310.

Gonçalves GA, Paiva RD. *Gene therapy: advances, challenges, and perspectives.* Einstein (Sao Paulo). 2017;15:369-75.

Gulati M, Bajad S, Singh S, Ferdous AJ, Singh M. Development of a liposomal amphotericin B formulation. *J Microencapsul* 1998;15(2):137-51

Hagstrom JE, Sebestyen MG, Budker V, Ludtke JJ, Fritz JD, Wolff JA. Complexes of non-cationic liposomes and histone H1 mediate efficient transfection of DNA without encapsulation. *Biochim Biophys Acta* 1996;1284(1):47-55.

Hofland H, Huang L. Inhibition of human ovarian carcinoma cell proliferation by the liposome-plasmid DNA complex. *Biochemical and biophysical research communications*. 1995;207(2):492-6.

Hui SW, Zhao YL. Mechanism of DNA uptake of transfection mediated by cationic liposomes. *Zoological studies*. 1995;34:73-5.

Hyde SC, Gill DR, Higgins CF, Trezise AE, MacVinish LJ, Cuthbert AW, Ratcliff R, Evans MJ, Colledge WH. Correction of the ion transport defect in cystic fibrosis transgenic mice by gene therapy. *Nature*. 1993;362(6417):250-5.

Ibáez M, Gariglio P, Chávez P, Santiago R, Wong C, Baeza I. Spermidine-condensed DNA and cone-shaped lipids improve the delivery and expression of exogenous DNA transfer by liposomes. *Biochem Cell Biol* 1996;74(5):633-43.

Ibraheem D, Elaissari A, Fessi H. Gene therapy and DNA delivery systems. *International Journal of pharmaceutics*. 2014;459(1-2):70-83.

Innes CL, Smith PB, Langenbach R, Tindall KR, Boone LR. Cationic liposomes (lipofectin) mediate retroviral infection in the absence of specific receptors. *Journal of Virology*. 1990;64(2):957-61.

Kaneda Y, Iwai K, Uchida T. Introduction and expression of the human insulin gene in adult rat liver. *Journal of Biological Chemistry*. 1989;264(21):12126-9.

Kaneda Y, Morishita R, Tomita N. Increased expression of DNA co-introduced with nuclear protein in adult rat liver. *Journal of molecular medicine*. 1995;73: 289-97.

Kato K, Kaneda Y, Sakurai M, Nakanishi M, Okada Y. Direct injection of hepatitis B virus DNA into liver-induced hepatitis in adult rats. *Journal of Biological Chemistry*. 1991b;266(33):22071-4.

Kato K, Nakanishi M, Kaneda Y, Uchida T, Okada Y. Expression of hepatitis B virus surface antigen in adult rat liver. Co-introduction of DNA and nuclear protein by a simplified liposome method. *Journal of Biological Chemistry*. 1991a;266(6):3361-4.

Kharakoz DP, Khusainova RS, Gorelov AV, Dawson KA. Stoichiometry of the dipalmitoylphosphatidylcholine-DNA interaction in the presence of Ca2+: a temperature-scanning ultrasonic study. *FEBS Lett* 1999;446(1):27-9.

Khosravi-Darani, K., Mozafari, M. R., Rashidi, L., Mohammadi, M. Calcium-based nonviral gene delivery: an overview of methodology and applications. *Acta Med Iran*. 2010;48(3):133-141.

Kulkarni JA, Witzigmann D, Thomson SB, Chen S, Leavitt BR, Cullis PR, van der Meel R. Author Correction: The current landscape of nucleic acid therapeutics. *Nature nanotechnology*. 2021;16(7):841.

Lakkaraju A, Dubinsky JM, Low WC, Rahman YE. Neurones are protected from excitotoxic death by p53 antisense oligonucleotides delivered in anionic liposomes. *J Biol Chem* 2001;276(34):32000-7.

Lam, AM, Cullis, PR. Calcium enhances the transfection potency of plasmid DNA-cationic liposome complexes. *Biochim Biophys Acta* 2000;1463(2):279-90.

Lamb, RA. Fusion of the myxovirus: a hypothesis for changes. *Virology*. 1993;197(1):1-1.

Lappalainen, K., Jääskeläinen, I., Syrjänen, K., Urtti, A., Syrjänen, S. Comparison of cell proliferation and toxicity assays using two cationic liposomes. *Pharm Res* 1994;11(8):1127-31.

Lasic DD. *Liposomes in gene delivery*. CRC press; 2019 Jul 23.

Lechardeur D, Sohn KJ, Haardt M, Joshi PB, Monck M, Graham RW, Beatty B, Squire J, O'brodovich H, Lukacs GL. Metabolic instability of plasmid DNA in the cytosol: a potential barrier to gene transfer. *Gene therapy*. 1999;6(4):482-97.

Legendre JY, Szoka Jr FC. Delivery of plasmid DNA into mammalian cell lines using pH-sensitive liposomes: comparison with cationic liposomes. *Pharmaceutical research*. 1992;9:1235-42.

Leibiger B, Leibiger I, Sarrach D, Zühlke H. Expression of exogenous DNA in rat liver cells after liposome-mediated transfection *in vivo*. *Biochemical and biophysical research communications*. 1991;174(3):1223-31.

Leventis, R., Silvius, JR. Interactions of mammalian cells with lipid dispersions containing novel metabolisable cationic amphiphiles. *Biochimica et Biophysica Acta (BBA)-Biomembranes*. 1990;1023(1):124-32.

Li, AP, Myers, CA, Kaminski, DL. Gene transfer in primary cultures of human hepatocytes. *In Vitro Cellular & Developmental Biology-Animal*. 1992;28:373-5.

Li S, Huang L. *In vivo* gene transfer via intravenous administration of cationic lipid-protamine-DNA (LPD) complexes. *Gene Ther* 1997;4(9):891-900.

Lipsky NG, Pagano RE. Intracellular translocation of fluorescent sphingolipids in cultured fibroblasts: endogenously synthesised sphingomyelin and glucocerebroside analogues pass through the Golgi apparatus en route to the plasma membrane. *The Journal of Cell Biology*. 1985;100(1):27-34.

Litzinger DC, Huang L. Phosphatodylethanolamine liposomes: Drug delivery, gene transfer and immunodiagnostic applications. *Biochimica et Biophysica Acta (BBA)-Reviews on Biomembranes*. 1992;1113(2):201-27.

Liu Y, Liggitt D, Zhong W, Tu G, Gaensler K, Debs R. Intravenous Gene Delivery. *Journal of Biological Chemistry*. 1995 Oct 20;270(42):24864-70.

Logan JJ, Bebok Z, Walker LC, Peng S, Felgner PL, Siegal GP, Frizzell RA, Dong J, Howard M. Cationic lipids for reporter gene and CFTR transfer to the rat pulmonary epithelium. *Gene therapy*. 1995;2(1):38-49.

Lollo CP, Banaszczyk MG, Chiou HC. Obstacles and advances in nonviral gene delivery. *Current opinion in molecular therapeutics*. 2000;2(2):136-42.

Lu D, Benjamin R, Kim M, Conry RM, Curiel DT. Optimisation of methods to achieve mRNA-mediated transfection of tumour cells *in vitro* and *in vivo* employing cationic liposome vectors. *Cancer gene therapy*. 1994;1(4):245-52.

Maja L, Eljko K, Mateja P. Sustainable technologies for liposome preparation. *The Journal of Supercritical Fluids*. 2020;165:104984.

Mali S. Gene therapy delivery systems. *Indian Journal of human genetics*. 2013;19(1):3.

McManus JJ, Radler JO, Dawson KA. Does calcium turn a zwitterionic lipid cationic? *J Phys Chem B* 2003a;107(36):9869-75.

McManus JJ, Radler JO, Dawson KO. Phase behaviour of DPPC in a DNA-calcium-zwitterionic lipid complex studied by small-angle X-ray scattering. *Langmuir* 2003b;19(23):9630-7.

Mellman I, Fuchs R, Helenius A. Acidification of the endocytic and exocytic pathways. *Annual review of biochemistry*. 1986;55(1):663-700.

Misra S. Human gene therapy: A brief overview of the genetic revolution. *J Assoc Physicians India*. 2013;61(2):127-33.

Mizuguchi H, Nakagawa T, Nakanishi M, Imazu S, Nakagawa S, Mayumi T. Efficient gene transfer into mammalian cells using a fusogenic liposome. *Biochem Biophys Res Commun* 1996;218(1):402-7.

Morishita R, Gibbons GH, Ellison KE, Nakajima M, von der Leyen H, Zhang LU, Kaneda Y, Ogihara T, Dzau VJ. Intimal hyperplasia after vascular injury is inhibited by antisense cdk 2 kinase oligonucleotides. *The Journal of Clinical Investigation*. 1994;93(4):1458-64.

Morishita R. Single intraluminal delivery of antisense cdc 2 kinase and PCNA oligonucleotides results in chronic inhibition of neointimal hyperplasia. *Proc. Natl. Acad. Sci.*. 1993;8474:90.

Mozafari, M. R., Hasirci, V. Mechanism of calcium ion-induced multilamellar vesicle-DNA interaction. *J Microencapsul* 1998;15(1):55-65.

Mozafari, M. M., Zareie, MH, Piskin, E., Hasirci, V. Formation of supramolecular structures by negatively charged liposomes in the presence of nucleic acids and divalent cations. *Drug Deliv* 1998;5(2):135-41.

Nabel EG, Yang ZY, Plautz G, Forough R, Zhan X, Haudenschild CC, Maciag T, Nabel GJ. Recombinant fibroblast growth factor 1 promotes intimal hyperplasia and angiogenesis in arteries *in vivo*. *Nature*. 1993;362(6423):844-6.

Nabel GJ, Nabel EG, Yang ZY, Fox BA, Plautz GE, Gao X, Huang L, Shu S, Gordon D, Chang AE. Direct gene transfer with DNA-liposome complexes in melanoma: expression, biologic activity, and lack of toxicity in humans. *Proceedings of the National Academy of Sciences*. 1993;90(23):11307-11.

New RR. Influence of liposome characteristics on their properties and fate. In *Liposomes as Tools in basic research and industry*. CRC Press. 2017: 1-20.

Nsairat H, Al-Shaer W, Odeh F, Essawi E, Khater D, Al Bawab A, El-Tanani M, Awidi A, Mubarak MS. Recent advances in the use of liposomes for the delivery of nucleic acid-based therapeutics. *OpenNano*. 2023:100132.

Nsairat H, Khater D, Sayed U, Odeh F, Al Bawab A, Al-shaer W. Liposomes: Structure, composition, types, and clinical applications. *Heliyon*. 2022;8(5):e09394.

Oku N, Yamazaki Y, Matsuura M, Sugiyama M, Hasegawa M, Nango M. A novel nonviral gene transfer system, polycation liposomes. *Advanced drug delivery reviews*. 2001;52(3):209-18.

Patil SD, Rhodes DG. Influence of divalent cations on the conformation of phosphorothioate oligodeoxynucleotides: A circular dichroism study. *Nucleic Acids Res*. 2000;28(12):2439-45.

Perrie Y, Gregoriadis G. Liposome-entrapped plasmid DNA: Characterisation studies. *Biochim Biophys Acta* 2000;1475(2):125-32.

Plautz GE, Yang ZY, Wu BY, Gao X, Huang L, Nabel GJ. Immunotherapy of malignancy by *in vivo* gene transfer to tumors. *Proceedings of the National Academy of Sciences*. 1993;90(10):4645-9.

Consider KP. Vectors in gene therapy. In *An Introduction to Molecular Medicine and gene therapy.* Edited by Kresnia TF, John Wiley & sons Inc, Newyork, USA. 2000;77-112.

Pradhan B, Kumar N, Saha S, Roy A. Liposome: method of preparation, advantages, evaluation, and its application. *Journal of applied pharmaceutical research.* 2015;3(3):01-8.

Ramamoorth M, Narvekar A. Non-Viral Vectors in Gene Therapy-An Overview. *J. Clin. Diagn. Res.* 2015;9:1-6.

Rinaldi C, Wood MJ. Antisense oligonucleotides: The next frontier for treatment of neurological disorders. *Nature Reviews Neurology.* 2018;14(1):9-21.

Robinson, E., MacDonald, K. D., Slaughter, K., McKinney, M., Patel, S., Sun, C., and Sahay, G. Chemically modified mRNA restores chloride secretion in cystic fibrosis. *Molecular therapy.* 2018;26(8):2034-46.

Ropert C, Lavignon M, Dubernet C, Couvreur P, Malvy C. Oligonucleotides encapsulated in pH-sensitive liposomes are efficient toward the Friend retrovirus. *Biochemical and biophysical research communications.* 1992 Mar 16;183(2):879-85.

Ropert C, Malvy C, Couvreur P. Inhibition of the Friend retrovirus by antisense oligonucleotides encapsulated in liposomes: mechanism of action. *Pharmaceutical research.* 1993;10:1427-33.

Ropert C. Liposomes as a gene delivery system. *Brazilian Journal of medical and biological research.* 1999;32:163-9.

Saffari M, Moghimi HR, Dass CR. Barriers to liposomal gene delivery: from the application site to the target. *Iranian Journal of pharmaceutical research: IJPR.* 2016;15(Suppl):3-17.

San H, Yang ZY, Pompili VJ, Jaffe ML, Plautz GE, Xu L, Felgner JH, Wheeler CJ, Felgner PL, Gao X, Huang L. Safety and short-term toxicity of a novel cationic lipid formulation for human gene therapy. *Human gene therapy.* 1993;4(6):781-8.

Schaefer-Ridder M, Wang Y, Hofschneider PH. Liposomes as gene carriers: efficient transformation of mouse L cells by the thymidine kinase gene. *Science.* 1982;215(4529):166-8.

Schmid RM, Weidenbach H, Draenert GF, Lerch MM, Liptay S, Schorr J, Beckh KH, Adler G. Liposome-mediated *in vivo* gene transfer into different tissues of the gastrointestinal tract. *Zeitschrift fur Gastroenterologie.* 1994;32(12):665-70.

Schmid SL, Fuchs R, Male P, Mellman I. Two distinct subpopulations of endosomes involved in membrane recycling and transport to lysosomes. *Cell.* 1988;52(1):73-83.

Schmidt-Wolf GD, Schmidt-Wolf IG. Nonviral and hybrid vectors in human gene therapy: an update. *Trends in molecular medicine.* 2003;9(2):67-72.

Singer SJ, Nicolson GL. The Fluid Mosaic Model of the Structure of Cell Membranes: Cell membranes are viewed as two-dimensional solutions of orientated globular proteins and lipids. *Science.* 1972;175(4023):720-31.

Smith JG, Walzem RL, and German JB. Liposomes as agents of DNA transfer. *Biochimica et Biophysica Acta (BBA)-Reviews on Biomembranes.* 1993;1154(3-4):327-40.

Staedel C, Hua Z, Broker TR, Chow LT, Remy JS, Behr JP. High-efficiency transfection of primary human keratinocytes with positively charged lipopolyamine: DNA complexes. *Journal of Investigative Dermatology*. 1994;102(5):768-72.

Tardi PG, Boman NL, Cullis PR. Liposomal doxorubicin. *J Drug Target* 1996;4(3):129-40.

Templeton NS, Lasic DD, Frederik PM, Strey HH, Roberts DD, Pavlakis GN. Improved DNA: liposome complexes for increased systemic delivery and gene expression. *Nature biotechnology*. 1997;15(7):647-52.

Tomita N, Higaki J, Kaneda Y, Yu H, Morishita R, Mikami H, Ogihara T. Hypertensive rats produced by *in vivo* introduction of the human renin gene. *Circulation research.* 1993;73(5):898-905.

Torchilin, VP. Recent advances with liposomes as pharmaceutical carriers. *Nature reviews drug discovery*. 2005;4(2):145-60.

Wang, CY, Huang, L. Plasmid DNA adsorbed to pH-sensitive liposomes efficiently transforms the target cells. *Biochemical and biophysical research communications.* 1987;147(3):980-5.

Watanabe Y, Nomoto H, Takezawa R, Miyoshi N, Akaike T. Highly efficient transfection into primary cultured mouse hepatocytes by the use of cation-liposomes: an application for immunisation. *The Journal of Biochemistry*. 1994;116(6):1220-6.

Weiner, AL. Liposomes as carriers for polypeptides. *Advanced Drug Delivery Reviews.* 1989; 3(3):307-41.

Yanagihara I, Kaneda Y, Inui K, and Okada SA. Editor Dickson, G. *Molecular and cell biology of human gene therapeutics: Liposomes-mediated gene transfer.* Springer Science & Business Media; 2012; 64-82.

Yoshimura K, Rosenfeld MA, Nakamura H, Scherer EM, Pavirani A, Lecocq JP, Crystal RG. Expression of the human cystic fibrosis transmembrane conductance regulator gene in the mouse lung after *in vivo* intratracheal plasmid-mediated gene transfer. *Nucleic acid research.* 1992;20(12):3233-40.

Zabner J, Fasbender AJ, Moninger T, Poellinger KA, Welsh MJ. Cellular and molecular barriers to gene transfer by a cationic lipid. *Journal of Biological Chemistry*. 1995; 270(32):18997-9007.

Zareie MH, Mozafari MR, Hasirci V, Piskin E. Investigation of scanning tunnelling microscopy of liposome-DNA-Ca2+ complexes. *J Liposome Res* 1997; 7(4):491-502.

Zhou X, Klibanov AL, Huang L. Improved DNA encapsulation in pH-sensitive liposomes for transfection. *Journal of liposome research.* 1992; 2(1):125-39.

Zhu N, Liggitt D, Liu Y, Debs R. Systemic gene expression after intravenous DNA delivery into adult mice. *Science*. 1993; 261(5118):209-11.

About the Editors

Usama Ahmad

Dr. Usama Ahmad specializes in Pharmaceutics from Amity University, Lucknow Campus, India. He received his PhD from Integral University, India, in 2018. Currently, he is an Associate Professor of Pharmaceutics in the Faculty of Pharmacy, Integral University. From 2013 to 2014, he worked on a research project funded by the Science and Engineering Research Board-Department of Science and Technology (SERB-DST), Government of India. He has a rich publication record with more than 35 original journal articles, 3 edited books, 7 book chapters, and several scientific articles published in Ingredients South Asia Magazine and QualPharma Magazine. He is a member of the American Association for Cancer Research, the International Association for the Study of Lung Cancer, and the British Society for Nanomedicine. Dr. Ahmad's research focus is on the development of nanoformulations to facilitate the delivery of drugs that aim to provide practical solutions to current healthcare problems.

Anas Islam

Mr. Anas Islam, specializes in Pharmacology with a master's degree from Integral University Lucknow, currently holds the position of Lecturer in the Faculty of Pharmacy. Anas has made significant contributions to the field, evident in his publications of reviews in reputable journals and book chapters in renowned publishing houses. His research interests span various facets of pharmacology, including phytochemicals, drug delivery systems, nanotechnology, and exosomes. Anas's adeptness in scientific communication is showcased through his extensive publications and active participation in seminars and workshops. With a commitment to advancing pharmacological knowledge, his comprehensive skill set aligns seamlessly with the book's subject matter, ensuring a valuable editorial perspective.

Index

U

V